MW00786266

"When I first read Nathaniel Lin's Applied Business Analytics, I thought, 'I wish I had written this.' The points Nathaniel makes about analytics deciders hit a unique target—managers who today don't realize what it takes to drive business using data. Having 'analytics deciders' engaged at the 'intersections' makes it possible for innovation to flow throughout the organization, like traffic cops in downtown Manhattan. This is one of those books that you keep on your desk and refer to as you build an analytics team. Here are all the things I loved about this book: the Advanced Analytics Layer (it has immediate applicability); the case studies (which are practical); and the hands-on approach with KNIME (the real-world experience)."

—**Theresa Kushner,** Vice President Enterprise Information Management, VMware and former Sr. Director of Customer Intelligence, Cisco

"Nathaniel has done a terrific job of capturing the key components needed to build high-performing analytics to drive business growth. This is a 'must read' for senior executives involved in analytics for decision-making."

—**John F Carter,** Senior Vice President, Analytics, Insight & Loyalty, Charles Schwab

"Making analytics mainstream: There are scores of books on the application of analytics to organizations, however, Applied Business Analytics is unique in two respects. First, it identifies why organizations fail to truly leverage analytics to achieve transformational outcomes. Second, it proposes a framework in which business processes are modified to accommodate 'analytics' through a continuous dialogue, rather than isolated analytical experiments. Forward-looking business leaders who want to truly leverage Big Data and analytics to transform and lead their organizations would do well to take to heart the core message of this book."

—**Prat Moghe,** CEO of Cazena and former Senior Vice President of IBM Netezza

"Business analytics has developed into an area of intense focus for many organizations, yet its successful application remains more art than science. Nathaniel Lin's Applied Business Analytics represents a significant step forward in remedying that situation by enabling business leaders to create the teams and business processes required to succeed with analytics. Business analytics is by its nature a deeply technical topic, and Dr. Lin does not skimp on the technical details, including meaty yet approachable hands-on exercises. Applied Business Analytics keeps the technical discussion where it belongs: in clear service of well-articulated and relevant business problems that provide the necessary context for analytics projects to create meaningful value. No mere how-to guide, Dr. Lin's book provides a complete picture of the analytical organization, as well as the conditions and processes required to develop and enable analytics deciders, the data-driven change agents who propel organizations to move from gut-based decision making to rational, evidence–based decision making. Packed with numerous real-world case studies and war stories, Nathaniel Lin's Applied Business Analytics is a solid starting point for any executive seeking to transform an organization with analytics."

—**Adam Ferrari,** CTO Crisply and Strategic Advisor Big Data and Vice President of Product Development, Oracle, Former CTO, Endeca Technologies

"Leaders who are establishing a data-driven culture will benefit from the lucid analytical process and methodologies that Nathaniel demonstrates in Applied Business Analytics. The chapter about Analytics Eco-system includes a fantastic comparison of different team structures that exist today."

—**Jane Chen,** Head of Analytics Center of Excellence, Analog Devices

"I am positively impressed by Nathaniel's book. This book has a unique position in the marketplace, standing out as an insightful perspective balancing an 'evangelist book' and 'technical how-to cookbook.' I appreciate Nathaniel's thought leadership and guidance. "

"Applied Business Analytics is 70% business and technical context discussion and 30% of how to be effective in organizational framework. The most challenging part of analytics leaders in today's world is still the conflict or tension between business drivers and technical details. That is why 'analytics deciders' are critical for maximizing analytics' functional impact on businesses. Balancing 'technical details (how)' with 'why analytics is critical to your success' and 'what result you can (expect to) achieve by building analytics capabilities' is challenging. Nathaniel's personal stories, case studies, and real-world success are helpful. I believe this book can be a great reference guide for emerging analytics leaders."

—**Gary Cao,** Vice President, Information Services & Analytics Strategy, Cardinal Health, Inc.

Applied Business
Analytics

Applied Business Analytics

Integrating Business Process, Big Data, and Advanced Analytics

Nathaniel Lin

Editor-in-Chief: Amy Neidlinger
Executive Editor: Jeanne Glasser Levine
Operations Specialist: Jodi Kemper
Cover Designer: Chuti Prasertsith
Managing Editor: Kristy Hart
Project Editor: Deadline Driven Publishing
Copy Editor: Nancy Sixsmith
Proofreader: Deadline Driven Publishing
Indexer: Angie Martin
Compositor: Jake McFarland
Manufacturing Buyer: Dan Uhrig

© 2015 by Nathaniel Lin
Upper Saddle River, New Jersey 07458

For information about buying this title in bulk quantities, or for special sales opportunities (which may include electronic versions; custom cover designs; and content particular to your business, training goals, marketing focus, or branding interests), please contact our corporate sales department at corpsales@pearsoned.com or (800) 382-3419.

For government sales inquiries, please contact governmentsales@pearsoned.com.

For questions about sales outside the U.S., please contact international@pearsoned.com.

Company and product names mentioned herein are the trademarks or registered trademarks of their respective owners.

Printed in the United States of America

First Printing December

ISBN-10: 0-13-348150-6
ISBN-13: 978-0-13-348150-1

Pearson Education LTD.
Pearson Education Australia PTY, Limited.
Pearson Education Singapore, Pte. Ltd.
Pearson Education Asia, Ltd.
Pearson Education Canada, Ltd.
Pearson Educación de Mexico, S.A. de C.V.
Pearson Education—Japan
Pearson Education Malaysia, Pte. Ltd.

Library of Congress Control Number: 2014952168

To my parents Tah-Fu and Paw-Chu Lin, who have spent years encouraging me to do something special. Even though this book is nothing special, it is a token of my appreciation for their love and nurturing since they brought me into the world 60 years ago. Last, to the one who really matters, I would use the same dedication from my PhD thesis:

"...Jehovah, who stretches forth the heavens and lays the foundations of the earth and forms the spirit of man within him." —Zechariah 12:1

Contents

Foreword

This book is well-timed for the current business analytics zeit-geist—the spirit of the times—in a variety of ways. Analytics and big data have become accepted tools for business strategy and competition. In surveys, virtually every manager agrees that they are important resources for business. If you are not a believer in data and analytics as guides to decision making, it is becoming embarrassing or politically incorrect to say so. This is all good news, at least from my perspective.

The only problem with this broad-scale consensus is that many managers and organizations still lack the skills and understanding to make analytics work for them. Lin's book is directed at just this issue. As he says, there have been a variety of books that attempt to get managers excited about analytics and textbooks that delve into the details of analytics without much sense of the business context. This book is a bridge between those extremes. It's for managers who already believe that analytics are important, but who don't yet know enough to practice them personally.

This transition is important and, indeed, necessary. In sophisticated companies, it is becoming difficult to draw the line anymore between analytics and the way work is done. Business is analytics, analytics is business. This book is designed to enable regular business people to create and consume analytics, even advanced ones. The detailed modeling and analysis of data may continue to be left up to professional "quants," but the framing, interpreting, and communicating of the analytical work will be a highly collaborative effort among "deciders," as Lin puts it, and quantitative experts.

Another way in which the book fits with the contemporary analytical mood is its focus on business and analytical processes. Lin correctly points out that analytics are increasingly being embedded within key business processes. In that sense, they are becoming somewhat invisible; as I write today, in fact, Gartner named "advanced, pervasive, and invisible analytics" as one of the "Top 10 Strategic Technology Trends for 2015." This invisibility makes the importance of management understanding even greater; if your business is heavily dependent on analytics in the underlying technology and process infrastructure, someone needs to constantly ask intelligent questions about underlying assumptions, methods, and outcomes.

To continue the process orientation, many of the topics and insights in this book are structured in terms of an analytics process that ranges from data preparation to data analysis and interpretation. The idea of a process fits the current analytics environment well. For too long, analytics have been pursued as a slow, one-off, painstaking process that would be difficult to institutionalize. Today, however, many companies are beginning to create "analytical factories" that emphasize speed, scale, and repeatability. The means to do this is to view analytics as a process with clear steps, handoffs, and modules. That is the focus in the book, and it's buttressed by examples using the open source analytics tool KNIME, which is itself structured in terms of processes. If we adopt Lin's model of thinking about analytics in process terms, our analytics will have higher productivity and impact.

A final linkage between this book and the current set of analytics themes is the focus on both big data and small. Most current books are on traditional, small data analytics or big data, but few cross that chasm to address both. Lin moves comfortably between analytics on traditional data (for example, marketing propensity models) and such big data types as social media and text from online product reviews. It's clear to me that before long, we won't use the term "big data," but we will certainly want to deal with unstructured, large-volume data. In any case, you'll be well-prepared for a world that integrates and analyzes all types of data if you read this book. Lin helpfully supplies a list of likely data elements that an organization might employ for typical analyses in different business areas.

This book is not only up-to-the-moment, but extremely comprehensive. There is scarcely an area of business analytics that is not discussed. You may want to read it quickly when you acquire it, and then use it as a reference book as you encounter a variety of analytics situations over the next several years. The ideas are both classical and contemporary, and they won't go out of fashion or relevance for a long time.

Thomas H. Davenport,
Distinguished Professor, Babson College
Author of *Competing on Analytics and Big Data @ Work*

Acknowledgments

I would like to thank the following people: My wife, Lu, for her confidence and encouragement. Alejandro Simkievich, a good friend and fellow Sloanie, who wrote Chapter 6. Any errors or omissions are solely mine. I want to thank also *FT Times*, and Victor for reading the draft and providing valuable suggestions.

About the Author

Dr. Nathaniel Lin is a recognized leader in marketing and business analytics across various industries worldwide. He has over 20 years of frontline experience applying actionable advanced analytics strategies to the world's largest companies in technology, finance, automotive, telecommunications, retail, and across many other businesses, including IBM, Fidelity Investments, OgilvyOne, and Aspen Marketing Analytics.

Nathaniel is currently the Chief Customer Insights Officer of Attract China. He is leading the efforts to develop leading edge Big Data Analytics technology and knowledge assets to deliver unparalleled values to Chinese travelers and U.S. clients. Nathaniel is widely recognized as an expert, teacher, author, and hands-on leader and senior executive in the application of data and advanced analytics in a wide variety of businesses. He is also the Founder and President of Analytics Consult, LLC (www.analyticsconsult.com). He leverages his rich and unique expertise in business analytics to help companies optimize their customer, marketing, and sales strategies. Together with his team, Nathaniel serves as a trusted strategic advisor to senior management teams. He is frequently invited as the keynote speaker in analytics events and advised over 150 CEOs in the U.S. and aboard on analytics and Big Data issues. He was invited by WWW2010 as one of the four expert panelists (together with the heads of Google Analytics, eBay Analytics, and Web Analytics Association) on the Future of Predictive Analytics.

As a recognized analytics expert, Nathaniel has partnered with Professor Tom Davenport to benchmark analytics competencies of major corporations across different industries. He also demonstrates his passion in cultivating future analytics leaders by teaching Strategic CRM and Advanced Business Analytics for MBA students at the Georgia Tech College of Management, Boston College Carroll School of Management, and Quant III Advanced Business Analytics at Bentley University.

Nathaniel holds a PhD in Engineering from Birmingham University (UK) and an MBA from MIT Sloan School of Management.

Preface

I have to admit that this book turned out to be much harder to write than I had envisioned. Having spent decades applying analytics and building and leading analytics teams, I thought it would be simply a summary of what I had learned and done before. Unfortunately, I find myself constantly struggling to find the "Goldilocks" approach—am I covering too much technical detail or too little? The latter, in my opinion, is worse because the lack of details might prevent business readers from doing the hands-on analytics exercises. However, the former with too many technical details would likely discourage business readers from reading the book altogether!

After several rounds of revisions, I hope I have managed to strike a balance to ensure the original objective of writing an analytics hands-on book for business readers has been met. However, in the event you find the content either too technical or not sufficiently technical, I beg for your indulgence. My advice is that you simply skip over the offending sections and go onto the next section. I am sure there are enough examples and business cases in this book that any serious business reader aspiring to be analytical would find relevant and helpful.

Why Another Book on Analytics?

Despite the many titles on analytics, a quick survey reveals that most broadly fall into two categories. One group focuses on promoting how businesses benefit from analytics. These are usually written by "evangelists" promoting the virtues of analytics. At the opposite end are those books written by "quants" for other analytics "quants." These contain too much technical detail for business readers. Few books are written to bridge the gap. There is a need for a book with hands-on applications of business analytics that solve real business problems that business readers can understand, practice, and use. I often hear the laments of business leaders: "I am sold after reading books promoting the power of analytics, but *none can show me how it is done and how to actually apply it in my business* without requiring me to go back to school and getting an advanced degrees in analytics and statistics!" If you share the same sentiments, then this book is for you.

How This Book Is Organized

Before "getting your hands dirty" with hands-on exercises, this book devotes the first four chapters to the basics, that is, the data, analytics tools (from simple to leading edge), and their relationships to the various business processes. The book continues with how to embed analytics and integrate the findings in existing or new strategic processes. With this in mind, the chapters are organized as follows:

- Chapter 1, "Introduction," is an introduction to analytics. It shows how analytics as the refining of data will transform business the way refining of oil transformed the twentieth century. It also addresses definitions, potential areas of applications, some vital lessons learned, how analytics are positioned in the business context, and how analytics are integrated into the business process.

- Chapter 2, "Know Your Ingredients—Data Big and Small," discusses everything you need to know about the basic ingredients of successful business analytics. It defines the differences between Big Data and the various types of data that may be useful to business analytics. It also describes data formats and the increasing importance of poorly or unstructured data to business insights and how to handle them.

- Chapter 3, "Data Management—Integration, Data Quality, and Governance," provides everything you need to manage and analyze data, from processing diverse data, to the importance of governance on data security and privacy, to good data versus bad data, to quality and latency, and to where to go beyond internal data sources.

- Chapter 4, "Handle the Tools: Analytics Methodology and Tools," discusses the most commonly used analytics models and tools, and provides hands-on exercises for some of the most commonly used analytics modeling techniques so you can see and feel how they work within real business cases. Its coverage spans basic models such as regressions, clustering, decision trees to the more sophisticated Big Data analytics such as text mining, and sentiment analysis on real data including online reviews on Amazon Fine Food and TripAdvisor hotels. A value prediction methodology using Ensemble Regression Tree model is also presented.

- Chapter 5, "Analytics Decision-Making Process and the Analytics Deciders," elucidates the differences between the conventional and the analytics decisioning process. It starts by exploring how conventional decision processes often falls short. It also describes the analytics decisioning process, known in this book as the BAP (Business Analytics Process), and how the right "analytics deciders" are needed to work in the "intersection" between silos to ensure the BAP's success. It ends with notes on how to become, identify, recruit, and retain analytics deciders.

- Chapter 6, "Business Processes and Analytics," discusses the business processes that can benefit from analytics. It also covers today's ERP (Enterprise Resource Planning) that is fueled by analytics. It includes CRM (Customer Relationship Management), SCM (Supply Chain Management), Financial Management, PLM (Product Cycle Management), and Human Capital Management.

- Chapter 7, "Identifying Business Opportunities by Recognizing Patterns," explains why the current way of dividing customers into "segments" just won't cut it. It tells you how to leverage advanced analytics tools to detect patterns of behaviors, events, trends, topics, preferences, and other critical factors that can impact business. Finally, it discusses pattern detection and how the group behavior of customers may help marketers plan more effective engagements using insights from the Market Basket Analysis and customer segment persona discovered using analytics.

- Chapter 8, "Knowing the 'Unknowable,'" discusses those topics and questions that are often deemed unknowable or unique (UU). These may involve little or limited data, predicting individual behavior in real time, and determining causality and lever settings for complex situations. This chapter provides real-life business cases and actual analytics workflows to illustrate how enlightened businesses can know and leverage analytics in these apparent "unknowable" events, thus, gaining a competitive edge over the competition. A novel methodology is also described to predict customer wallets and how to use that to form effective strategies.

- Chapter 9, "Demonstration of Business Analytics Workflows—Analytics Enterprise," uses the proper ingredients and tools to answer the top business questions except CRM (Customer Relationship Management), which is discussed in Chapter 10. You are encouraged to download and modify the workflows so that you can try to run them and adapt them to your own needs.

- Chapter 10, "Demonstration of Business Analytics Workflows—Analytics CRM," describes how to use the proper ingredients and tools to answer the top business questions surrounding management of customer relations. It provides tips on how to intimately know the customers—who they are, their values, profiles, persona, where and how to find them, and their needs and wants behavior—all through the use of analytics. Important insights such as customer purchase preferences and patterns, loyalty and impending churn, and how new customers may be acquired cost effectively and won back using today's social, mobile, and other media. You are again encouraged to fill out the BAP process involving the analytics modules and to download relevant workflows to try them on your business with your own data.

- Chapter 11, "Analytics Competencies and Ecosystem," discusses your current competency, how to move to the next level, and the importance of a healthy analytics ecosystem. It also provides an example of an enterprise analytics strategic plan, including organization structure, talent management, roles and responsibilities, and tips on how to run an effective analytics organization and to leverage external partners in the analytics ecosystem.

- Chapter 12, "Conclusions and Now What?" provides a summary of learning and suggestions for the next steps you should take toward putting together a strategic roadmap for achieving analytics competency and maturity. A list of recommendations that readers can adopt to achieve analytics competencies in the various areas covered in this book is also included.

- The Appendix contains material that is important but may be too detailed for inclusion in the main body of the book. This includes the basics on the analytics tool KNIME and descriptions of some of the KNIME nodes. You are encouraged to

augment it by consulting the introductory texts on KNIME.
com and YouTube.

After Reading and Working Through This Book

After reading this book, you should be able to:

- Possess a sufficient understanding of the entire analytics process.
- Acquire analytics vocabulary and basic knowledge of how analytics models work.
- Explore business analytics models and be able to modify or build real-life sophisticated analytics models.
- Navigate across business and analytics processes. Drill down or aggregate up at any level of the workflow and follow the analytics process without gaps in data, results, and insights.
- Better communicate with the analytics team.
- Be a contributing member of a rapid prototyping analytics team to provide business directions on demand.
- Detect actionable business insights from seemingly insignificant analytics details ignored by data and model "crunchers."
- Ensure business focus and alignment of final deliverables, and ultimately be an effective analytics leader and decider.

Bon voyage!

1

Introduction

Raw Data, the New Oil

"Data is the new oil"[1] has become a common mantra among today's Big Data proponents. However, Clive Humby, the founder of dunnhumby (the well-known retail analytics arm of Tesco Supermarkets in the UK), has also said that "[data] unrefined cannot really be used." For this reason, many skeptics argue that Big Data is not new and is really just a fad. Before you are convinced of the importance or insignificance of Big Data, let's try to use the "data as oil" analogy and show you the remarkable impact oil has had on every aspect of life across the world and (by inference) data.

If you look at a photo taken in New York in the late nineteenth century, you don't see cars and buses. Instead, you see commuters on horseback or in horse-drawn carriages and omnibuses. In fact, New Yorkers living in a city with a million people in the late nineteenth century enjoyed their night life lighted not by using electric or gas lamps, but by burning whale oil. In fact, at its peak, more than 10 million gallons of whale oil per year[2] (see Figure 1-1) were consumed across the United States. Because this usage was one of the main causes of the near-extinction of whales, Samuel M. Kier may be said to have saved the whales by successfully refining the crude oil into usable fuels in 1851. Of course, he also helped the emergence of entrepreneurs such

1 http://ana.blogs.com/maestros/2006/11/data_is_the_new.html
2 http://archive.org/stream/historyofamerica00tow#page/126/mode/2up

1

as Rockefeller, who started out by selling kerosene as one of the first products of the oil-refining process.

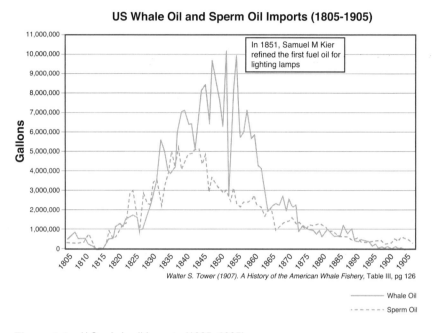

Figure 1-1   U.S. whale oil imports (1805–1905)

Over the next century, the impacts of oil literally reached the sky when U.S. astronauts landed on the moon in 1967. It is fair to say that oil transformed every facet of human culture and existence in the last 100 years. Just as you couldn't ask someone in the early 1900s to predict how oil would transform the world in 100 years, it is just as difficult to imagine today what a transformational role data and analytics might have in the rest of the twenty-first century.

Data Big and Small Is Not New

Although gasoline was new, crude oil was not "new" in the 1900s. In fact, crude oil has been known since prehistoric times. For 7,000 years, humans have been trying to find uses for crude oil.[3] In fact, one of its main uses in the United States in the nineteenth century was as folk medicine. Likewise, data (some may argue even Big [voluminous] Data) has been around for decades. What is new today is our ability to refine it to uncover novel insights and discover new applications.

In this book, I will show how business analytics (BA) is today's way of refining raw data, big or small, into strategic business opportunities. In many schematics for Hadoop and its popular Big Data-related tools such as Hive and others, these Big Data management tools are labeled the "Big-Data Refinery."[4] More appropriately, the Hadoop Distributed File System (HDFS) is more like an oil storage and distribution system than the actual refinery. The actual refining process happens right after the data has been properly stored and then subjected to a refining process, which is defined in this book as the applied business analytics process (BAP).

Definition of Analytics

At this juncture, we should define what the terms "analytics" and "business analytics" mean as they are used in this book. The standard dictionary definition of analytics, which is "the method of logical analysis" with a first-known use in 1590, clearly needs updating. I have also seen that the understanding of analytics in some businesses have not progressed much further. During an interview in 2010 of senior business executives with Professor Tom Davenport in benchmarking analytics competencies of major companies, I was shocked to hear one SVP of analytics consulting asserting that "Everyone in my team is doing analytics as they all work with numbers." Unfortunately, "working with numbers" does not make one an analytics practitioner.

---

3 http://ocw.mit.edu/courses/political-science/17-906-reading-seminar-in-social-science-the-geopolitics-and-geoeconomics-of-global energy spring-2007/lecture-notes/lec02_02152007.pdf
4 http://hortonworks.com/blog/big-data-refinery-fuels-next-generation-data-architecture/

A horse carriage has four wheels just like a car, but it is clearly not a car. We did not find any evidence that any "analytics consultants" were doing any analytics as defined here.

In this book, I define "analytics" (or "business analytics") this way:

- **More than just numbers**—Analytics is more than just working with numbers and data to find and report observed correlations and/or statistical distributions.

- **Knowledge and results-centric**—Business analytics is focused on the process of discovery of actionable knowledge and the creation of new business opportunities from such knowledge.

- **Tools-agnostic**—Analytics can use any computation or visualization tools from statistics, computer science, machine learning, and operational research to *describe* and *recognize* patterns, *seek* and *validate* causal relationships and trends, and *predict* and *optimize* outcomes.

To refine data into "business solutions" using the oil analogy, today's analytics "Rockefellers" must be able to do the following:

- **Understand the data**—Know the properties and nature of crude oil.

- **Understand the technology**—Know the different ways of refining crude oil.

- **Understand needs**—Know the human needs that are currently either barely met or unmet by today's technology.

- **Be creative**—Possess creativity in linking newfound properties of refined crude oil to meet known needs, and even create new markets out of new needs such as synthetics in materials, cosmetics, and pharmaceuticals which did not exist before.

Let's take a look at some of these business opportunities for applying analytics.

Top 10 Business Questions for Analytics

In my more than 15 years of direct involvement in applying analytics to business, I have found many important and difficult business questions that only analytics can answer. Today, the list is growing longer as more companies are investing in and applying analytics to new problems.

Although some of these questions can be answered from simply "slicing and dicing" raw data, some are so novel that many people in business might label them as "unknowable." I will give you one such example; you can then come up with your own list of unknowable questions to see whether analytics can help answer them.

One example is to predict the actual wallets of individual customers—and by extension the business share of their wallets. It was commonly viewed by business as unknowable because you can see only the purchases your customers make at your store, not their purchases at other stores.

Other examples of unknowable questions include what online visitors do when they leave your sites, what your customers need when they can't even describe it to themselves, and what to ship to the customers even before they start to consider buying and ordering. The former is the basis of the current ad exchange networks, in which third-party cookies are exchanged to let websites know where the visitors may have been before coming to your site. The last example is actually what Amazon is contemplating launching soon.[5]

Let's look at the first example. With a partial view, the true wallet size each customer possesses is elusive and can be inferred only through surveys. However, such survey data is often unreliable and hard to extrapolate to other individual customers. However, as to be explained in more detail in Chapter 8, I have directly used the best

---

5 "Amazon says it can ship items before customers order." http://www.usatoday.com/story/money/business/2014/01/18/amazon-anticipates orders/4637895/

customers' transactions as "lenses" to predict wallets for both business-to-business (B2B) and business-to-customer (B2C) customers. These wallets were successfully tested and deployed in many business problems.

Here is my list of top 10 questions that I feel can best be answered by the use of analytics. The questions have been divided into four sections: Financial Management, Customer Management, HR Management, and Internal Operations.

How analytics can answer the questions might not be apparent at the moment. One of the aims of this book is that you will be able to either find the answers from the examples in this book or be able to customize your own BA workflows for specific industries or business situations. Once comfortable with modules and processes, you might want to go to the list of questions in Chapters 9 and 10 to practice putting together your own analytics workflows to answer the questions.

The following sections discuss my top 10 strategic questions and tips on how analytics can be applied to provide answers.

Financial Management

Are your business and financial goals the right ones?

For a business to set the right goals, it is best to start from the customers and the current market. Because the entire market is rarely addressable due to its product focus or competitive landscape, a business needs to define the markets in which it wants to play. Once decided, analytics can then be used to predict the size of the addressable market and combine with revenues to obtain the respective market share.

Based on customer-level analytics, a business can predict what investment levels are needed to reach specific goals. Again, a business should start at the most granular levels and aggregate up to generate higher-level (business unit [BU] and enterprise) views. Some of the costs may involve realignment of business and product offerings.

Where are the areas of major opportunities?

The conventional top-down approach to market sizing is often not divisible at lower geographic or customer group levels. It has been a formidable exercise, often rife with contention when it was time for the BUs to assign annual sales targets at the various units based on the current market size predicted by the marketing intelligence (MI) team (at IBM during my time there) using the top-down approach. It felt like a prefab house being cut up and retrofitted—often messy and not a pretty sight.

However, it is just the opposite when you start with the smallest building blocks. You can build anything provided that you have a plan, which is the case when the addressable market is produced from predicting individual customer wallets. Armed with the customers' current spends and propensities to spend more, a business can find ways to more appropriately engage and entice different customers to shop more. Summing the wallets over the appropriate business units, the business can size the expected opportunities and identify the major revenue opportunities.

Are your investments adequate?

When the opportunities are identified, the natural next question is whether you are spending enough to win. Before knowing whether the investments are adequate, it is important first to predict the effective "levers" for increasing product wallet shares among the different groups of customers. Once the model results are validated with holdouts, the incremental causal effects of the levers need to be tested in-situ. A properly designed, multifactorial test-and-learn experiment can help determine the optimal combination of levers to give the maximum impacts. With the levers known, the business can then ascertain the costs of implementing the levers.

With the costs and impacts of the various levers determined, the business can put together a pro forma simulator to model the level of investments needed to move the needle (that is, per-unit wallet share gain). Once the optimal return on investment (ROI) has been decided, the requisite costs then need to be checked against investments in the annual strategic plan to ensure that adequate funds have been allocated to achieve the goals.

Customer Management

Do we know enough about our customers?

The conventional way to know customers was to hear directly from the voice of the customers through focus groups and surveys. Unfortunately, focus group and survey results cannot be easily applied to each individual customer. They also take time to organize and are costly to run regularly.

Given the richness of today's customer data, analytics has been shown to be capable of generating detailed individual and actionable customer knowledge. This knowledge includes predicting individual customer's propensities to buy from your brand and product, how much customers buy from you and from your competitors, and how valuable customers are to your business over their lifetime. Analytics can also perform customer multidimensional segmentation (behavior, demographic, attitudinal, and value) and develop segment persona. From the persona, the business can then devise effective strategies to better serve the segment, improve customer experience, and increase customer satisfaction and loyalty.

Lapses can happen despite best efforts, so the business should constantly predict and monitor customers' propensity to churn. The business should try to understand why they churn, what went wrong, and how to fix the problem. Any fixes should also be tested with control groups to make sure they are the right options. During and after implementation, those customers with high churning propensities should be surveyed to ensure that their issues were successfully resolved.

All vital customer knowledge assets should be shared, managed, and leveraged for all marketing, sales, operations, product development, and finance initiatives.

What actionable customer insights do you have?

The chief marketing officer (CMO) of a well-known steakhouse chain client received a customer insight report taken from surveys from a top management consulting firm. It showed that the client has more than 15 million customers that like its food, but they rarely visit the restaurants. It was a great insight, but where and how can they reach and entice such customers to come and dine more frequently? They can deduce from surveys what these customers might look like and devise broadly targeted mass media ads to entice them. They can't be sure whether the strategy would work across their chain when they test at a particular region or city. There are simply too many things at work to determine the effectiveness with sufficient certainty.

To make such insights actionable with analytics, the individual diners' receipts should be linked to the diners—either through a frequent diner program or by their name, credit card, or checking account number. The customer insights developed for the previous question can then be used to formulate campaigns and strategies to test and validate the effectiveness of the solutions derived from analytics insights. With a properly designed test-and-learn methodology, the incremental effects of levers can be ascertained before full-scale rollout. This analytics-driven approach would save costs and also reduce risk of failure.

Are we focused on the right social and mobile issues?

Social and mobile uses and data should be viewed as part of the data analytics value chain. Social and mobile are just part of the media. The customers' social, online, and mobile behavior data must be linked to target behaviors in terms of purchases and conversions. Once integrated, the issues must be germane to the BA process, actionable, aligned with, and have significant impacts on the critical strategic goals.

Is the critical knowledge from analytics properly managed (that is, captured, stored, shared, and reused) as an enterprise asset?

As discussed in Chapter 4, analytics innovation and knowledge tend to occur at the key intersections in which different disciplines, roles, functions, and goals "collide." To ensure that such knowledge can flourish and be captured, stored, shared, and reused as an enterprise asset, a process such as BAP (also see Chapter 4) should be adopted. Beyond the process, key "analytics deciders" who will be defined later must be present at the intersections. For now, an analytics decider is defined as someone who is highly proficient in both business acumen and analytics. A knowledge management IT system and a functioning "analytics sandbox" support the analytics deciders with well-defined rules, ensuring a safe and collaborative environment for creative and productive collisions.

HR Management

Do you have the right strategy for recruiting, managing, and retaining analytics talent?

In my years spent in building and managing analytics teams for big and small companies, I rank organizational and personnel issues as the number-one cause of failure in applying analytics. Many efforts failed because the analytics had either a limiting or conflicting reporting structure; the wrong compositions of talents and skills in the business, data, modeling, and strategy functions; or simply the wrong people (nonanalytics deciders) in key leadership roles within the intersections. Alternatively, companies' analytics efforts failed as they tried to remove the collisions at the intersections by delineating clear lines of responsibility. Unknowingly, these intersections were eliminated as everyone tried hard to stay within their own areas of responsibilities and roles. (Chapter 11 addresses some of these organizational and personnel issues.)

Internal Operations

Are your business processes driven with insights from predictive customer analytics?

Although companies started to use more predictive customer analytics in the past decade, many did not progress beyond tactical campaign targeting. As a result, the effectiveness of analytics can deteriorate over time. Sadly, many pioneers in analytics applications have been stuck doing the same things or have relegated analytics to just one of the support functions. More progressive companies have quietly embedded predictive analytics in their various business processes, and their CEOs rely on analytics for answers to their strategic questions.

For example, during a benchmarking project of companies' analytics competency, I was told that the CEO of one of the major online retailers, while preparing the annual report to the Wall Street analysts, discovered that revenues went up but the number of visitors went down. He wondered whether he should pose it as a positive or negative indication. Instead of consulting with the CFO or anyone within his senior leadership, he instead asked the director of analytics to come up with an answer. The CEO was satisfied with the reply and used the analytics recommendations during the call with the analysts.

I firmly believe that in the next few years, companies will win or lose by how much they can integrate and embed business analytics into their various business processes (as explained in Chapter 6).

Are your sales efforts fueled by analytics insights?

One of my earliest successes in applying analytics was not with marketing efforts, even though my team was part of the IBM MI team; it was to help IBM sales efforts. Since then, I find that sales efforts can be greatly enhanced by analytics insights such as leads generation, leads prioritization, sales force optimization, and telesales/call center performance analytics.

Vital Lessons Learned

Despite the power of BA, my experience in applying analytics in many companies across businesses has its share of frustrations that taught me many vital lessons. These lessons hopefully will aid you as you implement the BA Process for the first time and hopefully avoid wasting valuable time, resources, and opportunities.

Use Analytics

According to a 2013 survey of approximately 500 CMOs,[6] spending in analytics is likely to increase by 66 percent over the next 3 years, yet only 30 percent of the companies surveyed are using analytics. This is a decrease of about 19 percent (from 37 percent) from the previous year! Professor Christine Moorman, who directed the CMO survey, said "I think the mistake that a lot of companies make is that they spend on marketing analytics, but they don't worry enough about how you use marketing analytics." I have witnessed the same phenomenon in many companies.

Reasons Why Analytics Are Not Used

The common causes of senior management's lack of attention to how analytics is used are these:

- **Lack of understanding and trust**—Advanced analytics works, but few executives know how to use it, and fewer have direct experience working with it. As a result, management cannot vouch for or articulate analytics values when tough decisions are to be made among senior executive teams.

- **Overreliance on what they know**—Executives rise through the ranks doing things they know best. Unfortunately, analytics is likely not one of them. So when competing priorities demand their attention, they tend to rely on what they know, not on analytics. When the stakes are high, they often fall back on something they have prior experience with, not analytics (because it represents uncertainty and risk).

- **Lack of strategic vision of analytics' uses**—The true effectiveness of analytics is not in tactical applications such as

---

6 http://cmosurvey.org/files/2013/02/The_CMO_Survey_Highlights_and_Insights_Feb-2013-Final2.pdf

targeting or reporting. But that is exactly what some companies have been stuck doing, sometimes for more than a decade. Strategic applications cut across organizational silos and demand changes and even transformations of business, which might mean higher risks. Few executives would risk their careers to champion for the use of analytics under such situations.

- **Lost in translation**—Definitions and vernacular often differ between the senior executives and their analytics team, so business focus tends to get lost when traversing through the layers of management and finally reaching the modelers. In the reverse direction, significant customer insights might be ignored or filtered as "bad results" or "results not pretty enough to show the boss" by the analysts or mid-level management who might not interact so are probably not in tune with the senior executives.

The only way to mitigate this, I believe, is to ensure that there is a tight linkage between business and analytics. Hopefully, by taking the executives through the analytics process and enabling them to see analytics at work, they will gain a better understanding and be able to trust and use analytics!

Linking Analytics to Business

Instead of viewing analytics as a tool or a method, this book focuses on analytics as a process.

Business Analytics Value Chain

Rather than focusing on a single stage of the analytics value chain shown in Figure 1-2, this book takes a holistic view on the entire value chain (or value "ring" in this case) for continuous value creation. The focus is on the final business outcomes and their continued improvements using scientific test-and-learn methodology. To achieve sustainable wins over time, it is important that the process be run in a continuous fashion (this is the BAP that will be described in more detail in Chapter 5). Here's a brief description of the five major components:

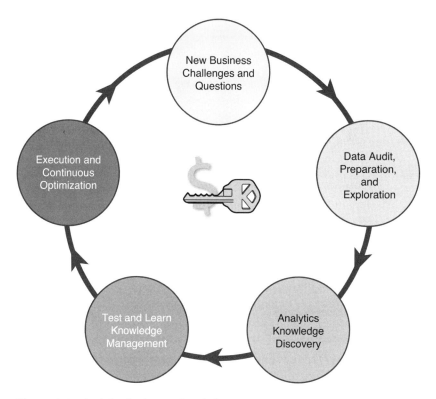

Figure 1-2   Analytics business value chain

- **Start with business questions**—All BA processes must start with a valid and high-value business question or idea. Even when the team is making a decision regarding IT, data, analytics models, and executions, it should not be made in isolation and apart from the business considerations. Even though the original business premise might sometimes need to be modified, it should always focus on business outcomes.

- **Conduct data audit and augmentation**—While the business objective is being set, a quick data audit can help determine whether the business has the right data to accomplish the objective. If not, additional data has to be acquired or derived to augment the existing data. Sometimes the costs of data augmentation can be too prohibitive in the beginning, and a scaled-back business objective can be created first to validate the idea and value before undertaking a full-scale project.

- **Extract knowledge**—The main goal of the analytics exercises, including simple business intelligence (BI) analysis and advanced analytics modeling, is to extract useful business insights, patterns, and knowledge from the data. This is usually an iterative process. By answering the initial question, a good analytics team often generates several more questions. For example, when you see a group of customers exhibit a certain behavior, your next questions may be these: Who are they? Are they high-value customers? How long has this been the case? Is this a recent phenomenon? By peeling the onion and answering the sequence of questions, you can reach the core of the problem and uncover something that may thus far be hidden and unknown.

- **Test insights and hypotheses and knowledge management**—Once an insight on the predictors' impacts on the business outcomes is obtained, it needs to be tested for causal effects. To do this, the scientific testing protocol known as the design of experiments (DOE) should be used with suitable control groups. No insights should be taken at face value, no matter how intuitive or elegant they are. Once validated, the insights and effects must be saved in a knowledge management system for sharing and reuse.

- **Execution and optimization**—Analytics insights must also take into account how they are to be implemented. The lever settings and their respective effects can be used to help optimize the analytics efforts. The validated insights and optimized lever settings can then be used to permit the business to ask the next level of questions. This begins another cycle of BAP.

Integrated Approach

Instead of viewing the various aspects of analytics techniques, business problems, and Big Data applications in isolation, I approach this by bringing everything together in the right sequence to produce the desired outcomes. This is why the book is not organized by analytics techniques, but instead by the way business problems are solved through analytics.

Hands-on Exercises

By embedding the entire analytics process with real business data within a powerful and modern advanced analytics platform (KNIME),[7] motivated business readers can then try real analytics on their own data to answer their question.

Note

For questions on installation and use of KNIME, please refer to the Appendix or the excellent KNIME booklet.[8] There is also a Learning Hub,[9] in which many great learning resources are available for all levels of users.

Reasons for Using KNIME Workflows

Based on the following reasons, I use KNIME workflows for all analytics examples and exercises in this book:

---

7 http://www.knime.com
8 http://www.knime.org/files/bl_sample_s.pdf
9 http://www.knime.org/learning-hub?src=knimeapp#basic

- **Advanced, powerful, and free**—As will be shown later, KNIME is not only powerful but (because it was developed recently) also incorporates the latest advanced analytics models, IT innovations, and user-friendly interfaces (for true click-and-drop analytics model development). Best of all, it is in the public domain and free (except for the server version).

- **Holistic and integrated views**—Some readers might want to see how the results are produced within an entire customer relationship management (CRM) or marketing or sales analytics process; others might want to drill down to a particular stage of the workflow to see how the data is transformed and/or model parameters are set and results produced. More adventurous readers might even want to vary the parameters and see how the changes affect the outcomes. The entire workflow can be viewed at the various levels of aggregation.

- **A single unifying enterprise workflow across silos**—In an ideal case, at the top level, it is strictly an enterprise business workflow. Below it lie the various sub-workflows that can correspond to the analytics workflows relating to IT, finance, marketing, stores, call centers, logistics, supply chain management, and so on. Drilling further down, you can then isolate the various components of the BA process. By embedding the actual analytics and reports within the metanodes of the sub-workflows, various stakeholders can then focus on the particular level and workflows they are interested in and be able to literally trace any questions to their sub-workflows to gain a holistic view of the big picture.

Conclusion

In this chapter, the transformational power of data, whether big or small, was discussed and likened to the way oil transformed the twentieth century. However, the power is not in data alone but also as it is being refined through analytics. Although analytics has become part of the daily vernacular, we have specifically defined BA as "the process of discovery of actionable knowledge and the creation of new business opportunities from such knowledge."

To illustrate the applications of analytics to answer many critical business questions, a list of top 10 questions in every area of business was discussed. Within each question, ways of applying analytics were also suggested. The vital lessons learned in more than a decade of applying analytics in businesses were also shared. To avoid the common trend of investing in analytics without using it, pitfalls were also highlighted based on the author's own experience.

The main lesson is to tightly link analytics to business. To do this, you have to create safe and productive intersections staffed by analytics deciders who are well-versed in both business and analytics, supported by an analytics sandbox as part of the holistic BAP.

2

Know Your Ingredients—Data Big and Small

"The most common cooking mistakes—you use inferior ingredients."

—CookingLight.com

Garbage in, Garbage out

With the increase in the use of business analytics (BA), the caveat "garbage in, garbage out" is more apt today than ever before. The proliferation of data has made it more imperative to pay attention to data before you attempt analytics. Because few analytics projects today are done with a single data type or source, there are questions about how data characteristics would affect the way data is to be integrated and used. To distinguish many new and less-structured data from the conventional data stored in databases, data-warehoused, and analyzed with Online Analytic Processing (OLAP) and business intelligence (BI) tools, the industry has coined a new term: "Big Data."

This chapter addresses the essentials of data and Big Data. Rather than treating them as two distinct types, we discuss data in a holistic manner, concentrating on the idiosyncrasies that are pertinent to business analytics (BA) and its impacts on business.

Data or Big Data

The term "Big Data" is a slight misnomer. The focus is not on the "Big" as in big volume, but on "Big" as indicating the severity of the challenges in handling and analyzing such data. There is no consensus on how severe these challenges should be before data becomes Big Data. The general agreement is that when data challenges start to impact business in one or more of the four "V" attributes defined in the following section, they become Big Data issues.

Definition of Big Data

In general, Big Data is data that is too big or too complex to be handled by standard relational database management system (RDBMS) technologies and has to be handled with additional technologies. Standard definitions for Big Data usually involve whether data possesses one or more of the following "V" key attributes:

- **Volume**—Data being handled is so voluminous that it frequently exceeds a server's storage and processing capacity. When vertical scalable solutions (adding more storage or faster processors) due to costs or zero downtimes are not acceptable options, horizontal scalable solutions (using cheaper servers without shutting down the existing server; for example, using MapReduce Hadoop technology) are needed.

  Business requirements for horizon scalability include the following:

  - Data grows too quickly over time and can overwhelm the capacity and processing power of the existing server.
  - Data needs to be stored first without a clear definition of relational data schema, or else the existing schema may not apply.
  - Data may be needed at all times and cannot be archived, or shutting down the server for maintenance and updating is not a viable option.

- Additional "cheaper" servers are needed for boosting the computational capabilities that can leverage the MapReduce Hadoop architecture.

- **Variety**—We are clearly facing a growing tidal wave of data from online, mobile, and social media; and from ubiquitous sensors. The sensors can be from the so-called Quantified Self (QS) personal devices, smart homes, and stores and linked devices as part of the Internet of Things (IoT). With new types of data, business data is no longer about well-structured relational data such as transactions, customer demographic data, and survey data. To handle diverse data, businesses must be able to store any data in a form that can be analyzed subsequently, which is the reason why a fourth "V" attribute ("veracity"—the lack of clarity or certainty) is added to the Big Data attribute list.

- **Velocity**—If the speed in terms of the data being generated, data being analyzed, and decisions being made is faster than the IT infrastructure can handle, you have a Big Data velocity challenge. The speed is often measured in fractions of a second as in real-time or how long it takes for a customer to click to leave your site or ignore your location-based mobile offer.

Data Types

Among the many types of data, we divide the data into four types: company data, consumer data, syndicated data, and other data (data that does not fit neatly into the other categories). One example is the aggregated sentiment data obtained through social media analytics.

The list of possible data is a long one and getting longer every day. Instead of an exhaustive list, I provide a generic checklist of the more important data for analytics and their characteristics and applications. You might want to add, remove, and customize the checklist for your specific business data needs and audits. Some data can be measured directly; others might need to be modeled and validated based on other data. Notice that the discussions are primarily in consumer data. Business-to-business (B2B) data shares many attributes, but is generally easier to handle than business-to-customer (B2C) data.

Company Data

As any company operates, much data is generated. This is usually kept for operational purposes and only until quarterly reports are generated. However, more and more companies realize that such data should be kept as long as possible. Company data can be a treasure store with potential insights waiting for discovery by BA. Let us examine this in detail in the following list:

- **Financial data (your company)**
  - ❏ Revenues
  - ❏ Costs
  - ❏ Cash flows
  - ❏ Margins
  - ❏ Targets and returns on investment (ROIs)
  - ❏ Trends of the preceding data
- **Point-of-sale (PoS) data**
  - ❏ SKU, unit costs, number of items bought, discounts and coupons used, loyalty card data, register number, store number
  - ❏ Coupons on checkout receipts
- **Marketing data**
  - ❏ **Channels**—With each passing day, there arise new channels that produce data and provide new ways to engage with customers. Here are a few examples: QS bands (Nike, Fitbit, Jawbone UP); IoT (UPC + unique serial item IDs+ real-time sensor data for automotive tires, tread, usage, wear pattern,

and tire pressure); social (Facebook, Twitter, LinkedIn, gaming communities, social communities on smartphone apps); and the social, local, and mobile (SoLoMo) omnichannels in which social, location, and mobile channels are leveraged together to ensure awareness and ultimately conversions.

Increasingly, some of the more complex online purchases, such as insurance policies and travel packages, require a live person for presale advice and handholding. Such services can best be provided through qualified call center reps. The data coming from the call centers can provide a rich source of data, not only on how well the reps perform but also the views, sentiments, and choices of customers.

To ensure that such data is properly captured, automated voice recognition software, text, and sentiment analytics and models to predict specific outcomes based on key insights mined automatically by the system must be customized and tested. Few companies have such capabilities today.

❏ **Campaign results**—Campaign performances, not just conversations but also critical sales lead stages beyond marketing, should be collected, stored, and analyzed.

❏ **Marketing returns on investment** (MROIs) of every marketing spending.

❏ **Promotions**—Promotions include promotion dates (month, day, and weekday), messages, offers, bundles, creative, and costs. More than a file or picture of the promotions and codes should be stored. To ensure that the information of the collaterals is query-able (that is, actual attributes in texts), metacodes describing the key features of the creative, offers, and so on should also be stored.

In my experience, companies spend a lot on external data, but often fail to characterize and store their own outbound communication and customer responses consistently. As a result, it is often hard to analytically model the optimal communications for a certain initiative because they would not know which of the features trigger what kind of response from the respective customer segments. Promotion-related data useful for analytics include these:

❑ **Advertisements**—Print, email, online, mobile, social, location, cable/TV, and events.

❑ **Targeting data**—Messages, offers, customer segments, tests done, response, and conversion rates.

❑ **Market research**—Surveys and focus groups on brands, sentiments, and competitive intelligence.

❑ **Offer bundles**—Bundles and discounts.

❑ **Spend**—Costs of material, printing, ads, channel, postage, agency, and internal resources.

❑ **Promotion collaterals**—Each campaign collateral should be consistently tagged with relevant keywords, descriptions, offer types, message types, individual customer responses per communication, and so on. Beyond reporting of campaign effectiveness, such data permits the optimization of models and campaign features using multivariate (design of experiments [DOE]) test-and-learn methodologies.

❑ **All internal promotion documents**—In terms of texts, pictures, and videos, offers should be "codified" and stored in NoSQL for subsequent analysis and as inputs to analytics projects. Few companies realize the importance of storing and analyzing their own communications with the subsequent customer responses. Although this kind of data can be thought of as regular surveys, it also reaches a wider audience and uses more reliable responses. Instead of verbal responses, the customers respond with their actions and purchases. Mining such a treasure trove of customer information can reveal valuable and timely insights.

It can be worthwhile for companies to follow what The Echo Nest does for music: to characterize the unstructured data into its most atomic attributes.[1] With these attributes, companies can then store, analyze, and optimize the contents of their communications. The same can be done for product attributes.

❏ **Channels**—With each passing day, new channels emerge that generate data and provide new ways to engage the customers. A few examples of channel data sources include the following:

❏ **Email**—Targeting lists, messages, subject line, offers, day of week mailed, open and responded, performance metrics and email performance metrics, and specific calls to actions and past results.

❏ **Web channel**—Web logs that can include date, time session ID, cookie-related data, clicks, referrer URL, referrer host, keyword or phrase search, and platform or operating system that the visitor came from. From these logs and user registrations, web analytics of hits, visits, unique visitors, visit length, path analysis, and abandoned shopping carts can be produced using various web analytics tools: Google and Yahoo Analytics, Coremetrics, Omniture, Webtrends, and so on.

❏ **Call centers**—Call centers are gaining more importance despite the fact that consumers are buying more products online. When making complex purchases such as buying car or life insurance and travel packages, consumers often prefer to talk to a live person for presale consultation.

---

1 http://sloanreview.mit.edu/article/what-people-want-and-how-to-predict-it/

Data from call centers can provide a rich trove of data on how well the reps perform and also on the views, sentiments, and choices of the customers. To make sure such data is captured, automated voice recognition software, text, and sentiment analytics to predict specific outcomes based on key insights mined automatically by the system must be customized and tested.

❑ **Personal health**—QS sensor bands that include Nike, Fitbit, Jawbone UP, and so on.

❑ **Internet of Things (IoT)**—Sensors are increasingly ubiquitous (for example, using UPC and real-time sensors to monitor and predict shelf product churn and yield). Other sensors can recognize customers and their movements throughout the stores.

❑ **Social**—Facebook, Twitter, LinkedIn, gaming communities, and social communities on smartphone apps.

❑ **SoLoMo channels**—These channels are leveraged together to ensure awareness and conversions.

❑ **Customer media usage and preferences**—Such data helps the planning of omnichannel strategy to offer different customers their preferred channels during their purchase life cycle stages.

❑ **Pricing**

  ❑ Pricing history

  ❑ Bundling

  ❑ Special promotions and their effects on revenues, customer experience, and loyalty

❑ **Products**

  ❑ Product hierarchy

  ❑ Suppliers

- ❏ Pricing
- ❏ Keywords and descriptions
- ❏ Versions and dates
- ❏ Inventory churn
- ❏ Category assortments and retail layout

- **Sales and services data**

❏ **Call centers**

- ❏ Sales management
- ❏ Performance metrics
- ❏ Routing history
- ❏ Costs per acquisition
- ❏ Costs per retention of churning customers
- ❏ Costs per recovery of dormant customers

❏ **Live chats and face-to-face reps**

- ❏ Text records of dialogues
- ❏ Duration
- ❏ Customer satisfaction
- ❏ Issues raised
- ❏ Issue resolution
- ❏ Customer experience
- ❏ Historical data

❏ **Event promotions**

- ❏ Types of events
- ❏ Numbers planned and attended
- ❏ Targeting criteria and models used
- ❏ Event performance metrics (attendance vs. value achieved)
- ❏ Media used for invitations

❏ **Service tiers**—Which customers should receive priority services instead of standard services

❏ **Warranty and after-sale services**

    ❏ Frequencies of repairs and inquiries

    ❏ Types of issues

    ❏ Customers segments

    ❏ Products

    ❏ Time to resolution

    ❏ Customer experience

    ❏ Customer sentiments and trend

- **Operational data**

❏ **Inventories and logistics; supply chain management (SCM)-related data**

    ❏ Number of items ordered, stored, and delivered per day

    ❏ Product churn durations

    ❏ Matching of products to customers near the distribution centers

    ❏ Aggregated demands (from purchases and recommendations)

    ❏ Shortages and overstocked items

    ❏ Historical performance metrics

❏ **Stores**

    ❏ Location and size

    ❏ Type of store (which store cluster)

    ❏ Customer compositions

    ❏ Neighborhood demographics

    ❏ Layout

    ❏ Assortments

    ❏ Displays

❏ Traffics

❏ Promotions

❏ In-store events

Individual Consumer Data

The importance of individual consumer data to business cannot be over-emphasized. Most of the data is from your customers, although some may be from online visitors to your sites. Many famous brands today, such as Amazon and Capital One, gained their competitive advantages because of their access to individual customer data. We discuss the individual customer or consumer data in the following list:

❏ **Personal identifiable information (PII)**—First and last name; mailing and billing addresses; mobile phone number; email addresses; customer ID, Facebook ID, Twitter ID.

❏ **Sensitive data**—For Social Security numbers (SSNs), credit cards, and other sensitive purchase data, the best practice is to not store such data for modeling, but to convert them into unique customer IDs. If the data has to be stored, it must be strongly encrypted and excluded from any analytical databases.

❏ **Demographics**—Age and gender, highest education.

❏ **Household information**—Information can be purchased from external vendors such as Acxiom and Experian, among many others. Here are some examples:

  ❏ Number of household members

  ❏ Assets

  ❏ Income

  ❏ Presence of teenagers, young children, babies, and elderly

  ❏ Presence of pets (cats, dogs, or others)

❑ **Customer values**

  ❑ Product purchases (past and predicted)

  ❑ Wallets per product and total addressable wallet

  ❑ Wallet share

  ❑ Lifetime value

  ❑ Historical trends

❑ **Customer behaviors**

❑ **Purchases**

  ❑ When, how many items, unit costs

  ❑ Discounts, use of coupons and promotions, bundled products purchased

  ❑ Loyalty points earned or redeemed

❑ **Online behavior**

  ❑ Pages, RSS, images, videos, blogs viewed

  ❑ Referrer URL, click stream, and browsing history

  ❑ Comments and blogs written

  ❑ Search terms

  ❑ Path analysis

  ❑ Page attribute analytics

  ❑ Abandoned carts

  ❑ Login dates and durations

  ❑ Last login

❑ **Mobile behavior**—Today's smartphones collect both passive and active data. Passive data is measured inherently by the phone (for example, from accelerometer, GPS, app usage, battery, 3/4G, Wi-Fi, compass, volume, disk space and usage, gyroscope, IP address, operating system, retina display, and service set identifier [SSID]). Active data is additional data collected by specific apps that can include email address, incoming and outgoing phone numbers, address book contacts, websites visited, key search terms, pictures, and so on.

❏ **Social media data**—Some of this data may be acquired from the respective application programming interfaces (APIs). However, with the tightening control from Facebook and LinkedIn, API access policies are constantly changing. It may be advantageous to store all the relevant social media data from your customers and prospects over time if you want to monitor trends of issues and sentiments. Another way is to engage with the various vendors, such as Gnip or Radian6, under Salesforce.com.[2] Forrester Research[3] recently evaluated some of the leading social marketing vendors.

❏ **Lifestyle data**

  ❏ Hobbies

  ❏ Magazine subscriptions

  ❏ Types of cars owned

  ❏ Pets

  ❏ Travels

  ❏ Cooking

❏ **Loyalty data**

  ❏ Loyalty cards

  ❏ Card number

  ❏ When joined

  ❏ Name and contact information

  ❏ Date last used

❏ **Loyalty tiers**

  ❏ Recency, frequency, monetary value (RFM) tier of usage

  ❏ Churning probability

---

2 http://www.salesforcemarketingcloud.com
3 "Social Advertising Platforms," Q4 2013 by Zachary Reiss-Davis, Forrester Research, December 17, 2013.

❏ Loyalty tier

❏ Shopper wallets and loan-to-value ratios (LTVs)

❏ Product wallets and values

Sensor Data

In addition to individual consumer or customer data, there are increasingly new data sources that are generated by the sensors that consumers wear or put in their properties such as homes, cars, and appliances. The confluence of all these data streams is likely to produce new business opportunities that are now unimaginable. Of course, with new great opportunities come great risks and greater responsibilities for businesses to be custodians and guardians of their customers' data. We will examine the various data sources as follows:

❏ **QS data**—Given the growing popularity of QS sensors such as the Nike Fuel Band, Fitbit, and Jawbone UP, such data may become more available to businesses in the near future. More customers are willing to sign in with their Facebook or Twitter ID and password. The types of personal data collected are usually activities, sleep patterns, diets, moods, and weight. (More features are being added as this book is being written.) Health insurance companies are beginning to incorporate such data into how they set premiums and rebates.

❏ **Smart home data**—With the availability of home-based sensors for climate and security controls, some of the data may offer ways to enrich current consumer data for more precise analytics. Again, the privacy and data security risks need to be properly addressed to avoid potential damages to the customer relationship and generate adverse public sentiments.

❏ **IoT**—Though talked about since the last century, the time has come for the practical application of IoT. Many gadgets and appliances can be connected as part of the "smart-*xxxx*" applications, where *xxxx* stands for *home*, *cars*, *city*, and so on (as in smart home, smart cars, smart roads, smart city, smart water, smart public transportation, smart air, smart medicine, and so on). A great overview of how a smart home can work is given by a recent Forrester Research report[4] and in the article, "Vision of Smart Home—The Role of Mobile in the Home of the Future."[5]

Much of the data collected by the connected sensors can be used in analytics to provide the "smartness" in all these smart applications. This data may also be used to aggregate values on behalf of consumers and to realize cost savings and new opportunities by more progressive businesses. These benefits can then be returned to the participating consumers as rewards for joining the mutually beneficial "commonwealth" of connected data. Without sounding cliché, we are truly at the cusp of a once-in-a-century opportunity of business transformation.

Syndicated Data

Most companies can enrich their internal customer data with data from external vendors. These vendors might include those that collect and model data from panels, surveys, and census data (such as Experian, Acxiom, Nielsen MRI,[6] and IXI). Newer vendors such as Gnip and Rapleaf provide the social media data and online behavior data based on email addresses. Gnip is the largest vendor of social media data collected from public APIs of sources such as Twitter and Tumblr, as well as managed integration from the public APIs of Facebook, YouTube, Google+, and others. Rapleaf claims to have

---

4 "The Internet of Things Comes Home, Bit by Bit," by Frank E. Gillett, Forrester Research Report, December 23, 2013.

5 http://www.gsma.com/connectedliving/gsma-vision-of-smart-home-report/

6 http://www.nielsen.com/content/dam/corporate/us/en/newswire/uploads/2009/05/mri_fusion_r2-26.pdf

collected real-time behaviors and profiles of 80 percent of consumers with email in the United States. With customer email, Rapleaf can provide additional fields to enrich the email data. For analytics, it is important to test the quality of the data by checking the incremental lifts provided by this additional data. If the data does not have any effect on the model results, it is best to leave it out because it might add to the noise and impede the effectiveness of the eventual models.

Syndicated data can be used in two ways: to profile customers and to assist in predicting the model outcome as additional independent variables (IVs) or predictors:

- **Profiling**—The syndicated data is useful to add color and appreciation of different customer persona and is checked against common customer perceptions. This process helps the marketing team come up with possible "levers" to elicit customer preferences and desired behaviors. In addition, the syndicated data from Experian, Acxiom, Nielsen Prizm NE, and P$YCLE Segmentation[7] can be used to enhance the model performance.

- **Model enhancement**—To enhance individual customer data, the internal transaction, loyalty, and behavior data can be augmented as follows:

  - Generate a random sample of target customers with their names, addresses, email addresses, and telephone numbers.

  - Send to the external data vendor for test append (it gives an indication of how well the vendor has coverage of your customers).

  - Select the appropriate data packages with the desirable attributes (including the vendor's own segmentation and modeled data). Balance the requirements between quality, coverage, and how informative.

---

7 http://www.claritas.com/sitereports/reports/psycle-demographics-reports.jsp

- If you are satisfied with the coverage and quality, send a suitably larger sample (1 percent or 5 percent sample) from your target customers.

- If you are modeling against noncustomers, ask for a suitably sized random sample from the vendor's own database to serve as the noncustomers for the model to differentiate them from the target customers.

- By testing the models with and without the additional IVs and using the appropriate variable reduction techniques,[8] you can then determine the incremental lifts of a subset of the vendor data. By reducing the number of variables to no more than a few dozen, you can avoid the problem of overfitting[9] (see Figure 2-1). Overfitting generally occurs when you have too many variables to fit an equation to the available data.

Increasing the number of predictor variables usually reduces training errors. Initially, the validation error decreases with more predictors (IVs). However, as the number of variables increases beyond a certain point, the validation error starts to increase: As the model tries to fit the training data set with more degrees of freedom (IVs), it starts to fit the noise unique to the training data set. When the model is applied to the validation data set, those conditions no longer exist, so the model performs worse than with fewer degrees of freedom.

---

8 http://www.biostat.jhsph.edu/~iruczins/teaching/jf/ch10.pdf
9 http://people.duke.edu/~mababyak/papers/babyakregression.pdf

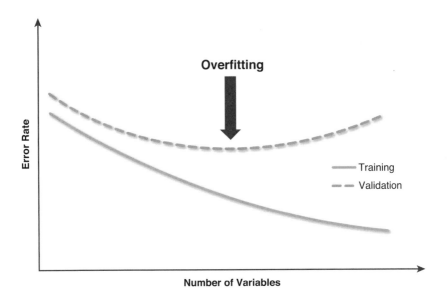

Figure 2-1   Example of overfitting

Data Formats

Text and numerical data can be transmitted in different formats. They can be in software-specific formats such as SPSS (.sav) and SAS (.sas7bdat), or more commonly in the various text formats delimited by commas (.csv), by tabs (.dat), or in an unspecified format (.txt). All text files can be read using the KNIME File Reader node. KNIME provides input nodes (see Figure 2-2) that can read many of these formats, including SAS data files. For data files stored in relational databases, KNIME can read them as long as they can be accessed via either Java Database Connectivity (JDBC) or Open Database Connectivity (OCDB) drivers. Even if you need to access data stored in an unsupported format, there are likely input nodes out there within the KNIME user community that do exactly what you want.

Figure 2-2   KNIME input reader notes options

Structured, Poorly Structured, and Unstructured Data

As the types and size of data set grow in today's business environment, it is important that all types of data are properly stored and ready for analysis, regardless of their size. The old days of just storing certain types of aggregated data (and only for 6 months) are long gone. Today's businesses have to store structured transaction data, high-frequency unstructured text, and nontext mobile behavior data and then integrate them for subsequent analytics. Because of various types of questions the business might need to answer, no prior relational schema can be specified *a priori*.

Traditionally, to ensure that database transactions were processed properly, a set of properties known as atomicity, consistency, isolation, and durability (ACID) had to be ensured (see Table 2-1). However, such constraints can be over-limiting and unnecessary today.

Table 2-1   Database ACID Compliance Definition

DB Properties	Description of Properties
Atomicity	All or none when updating database; prevents partial erroneous updates.
Consistency	All database transactions would not violate any integrity constraints at the start and end of the transaction; prevents loss of data integrity.
Isolation	Controls when and how the changes of an operation become visible to operations running at the same time; prevents errors caused by concurrent operations.
Durability	All "committed" transactions must survive permanently to prevent loss of committed data.

To ensure the capability to handle future data, requirements for horizontal scalability (that is, adding more storage and processing capacities without shutting down the system) are becoming inevitable. Table 2-2 describes and contrasts various types of databases. Given the available options, how do you leverage the strengths of the respective databases to meet different use cases? Despite the recent hype, not all data is suitable for storing on a Hadoop Distributed File System (HDFS).

Table 2-2   Overview of Database Systems

	RDBMS Relational DB	NoSQL	HDFS (Hadoop)	NewSQL (Google F1 Database)[12]
Horizontally scalable	✘	✔	✔	✔
ACID compliant	✔	✘	✘	✔
Can handle unstructured data	✘	✔	✔	✔
Distributed file system	✘	✔	✔	✔
Can handle large data set	✘	✔	✔	✔

Specialized batch analytics	✘	✘	✔	✘
Runs on inexpensive servers	✘	✔	✔	✘

Let's use an imaginary mobile music dynamic recommendation portal called *LikeaSong* as an example. Assume that the mobile portal has decided to recommend and serve music in real-time via ubiquitous smartphones. *LikeaSong* also wants to capture its users' listening, browsing, song contributing, and social behaviors for later analysis. To ensure that the users of the *LikeaSong* mobile app accept and like the recommended songs, advanced analytics is needed to link their behaviors and preferences to the characteristics of the songs and artists (from Gracenote[10] or The Echo Nest[11]). The approach that is adopted may look like the following:

- **Conventional data in SQL**—Customer data (registration and subscription) was stored for agile operational requirements on an SQL server.

- **NoSQL for real-time service**—To recommend and serve the music in real-time to listeners across the world, the only inexpensive but scalable solution was to store all the curated music, and to collect and store all user behaviors and preferences in either an HDFS or NoSQL database (refer to Table 2-2). Because Hadoop's specialty is batch processing, a NoSQL database was chosen instead.

- **Hadoop for Analytics**—To be able to integrate large data sets from music, mobile behaviors, and customer data, a Hadoop system was configured. By using Hadoop's MapReduce together with KNIME, the recommendation rules and curated

10 http://www.gracenote.com
11 http://echonest.com
12 http://static.googleusercontent.com/media/research.google.com/en/us/pubs/archive/41344.pdf

songs were constantly optimized to improve the users' experience and reliance on *LikeaSong* as their preferred portal for their music enjoyment.

One or more of the following conditions invariably render a conventional RDBMS undesirable and justify the use of NoSQL databases:

- **Data model issues**—Too complex, volatile, or uncertain.
- **Performance issues**—Scaling bottlenecks or cost escalation.
- **Integration and nonreporting analytics challenges**— Unable to handle either data integration from diverse sources and types, or real-time operational data requirements or the support of advanced analytics.

Conclusion

This chapter was an overview of various types of data. We discussed how Big Data might be defined and leveraged. The four Vs distinguish Big Data from small data and can impact analysis. Intimate knowledge can be inferred from customer data in terms of types, uses, and sources. New data generated by SoLoMo channels and sensors were discussed, and you learned how they can be collected, stored, and analyzed. Such knowledge is the fuel for novel business applications and innovations.

Before any analytics can be done, you should choose the right databases. This chapter also discussed the various criteria for choosing the right databases (such as RDBMS, HDFS, NoSQL, and NewSQL) to meet data and BA requirements.

3

Data Management—Integration, Data Quality, and Governance

"It is a capital mistake to theorize before one has data."

—Sherlock Holmes, *A Study in Scarlett* (Sir Arthur Conan Doyle)

Before any data can be used in business analytics (BA), some issues that need to be properly addressed and consistently managed include data integration, quality, security, privacy, governance, and data preparation and analysis. These data issues can determine whether a company's strategy in leveraging analytics succeeds or fails.

The power of today's data rarely resides in a single data source. The real insights come when multiple data from diverse sources can be integrated together. For example, a bank should be able to detect a potential attrition problem when a customer calls or texts the customer service rep in an online chat. To do this effectively, a business needs the following data to run analytics and formulate cost-effective intervention strategies:

- **Customer segment, persona, and value**—To inform who the caller is and determine the value and characteristics of the caller.

- **Propensity scores**—To indicate the likelihood of leaving the bank and to alert for intervention.

- **Social media data**—To identify critical emerging issues, to ensure that there is no systemic problem, and to pinpoint where the issues may be.

- **Rules and intervention options**—Based on all the previous data, to suggest intervention options, steps, and processes based on optimized and tested rules.

Any significant error or gaps in the integrated data uncorrected have the potential to doom the entire project.

Data Integration

To ensure that data from different sources can be merged at the most granular level, the following questions must be addressed first:

- **Identify data**—Is the data identifiable at the customer level?

- **Define unique key**—Is there a unique key (such as a customer unique ID, phone number, or email address) that links across the data sources, either directly or derived?

- **Bridge data gaps**—Is the entire data cycle a closed loop without any gaps, including the targeted behavior? If there are gaps, how can such gaps be bridged?

- **Assess data fits**—Are the various data sets accessible at the right time, customer, and product aggregation levels? For example, real-time web or mobile log data is not readily inferable with customer past transaction data unless we have at least the following:

  - A direct link between online users and the customer ID (cookies, login information, mobile number, IP address, or less precisely by their browsing pattern)

  - Data aggregated at the desirable time intervals (hourly, daily, or weekly)

- Indirect links; in the absence of a unique identifiable ID for a particular data set, you can use indirect means to link the data to any model results or other data set by using the following:
    - Customer demographics
    - Customer behaviors (what was bought, organic versus regular groceries, a certain size of designer clothing)
    - Similarity in terms of purchases or use of checks or credit cards to other households within a certain radius from a certain store

Data Quality

Quality is *relative*, especially for business data. In my decades of working on data, I have rarely (if ever) encountered a data set that does not have any deficiencies or errors. On the other hand, it is equally rare to find data so bad that it does not support any business analysis. Case in point: When I tried to apply advanced business-to-business (B2B) targeting analytics for IBM in China, I was told the data collected contained too many errors for any meaningful analytics. However, even with only 20 percent workable data for the models that were eventually built for the China market, the models could achieve lifts of more than 700 percent. To know whether a particular data set is of adequate quality, we need to define what "workable data quality" means.

Workable business data quality can be defined as data with sufficient quality to meet business needs in the following attributes:

- **Accuracy**—Data is free from significant errors.
- **Completeness**—Contains data from all requisite sources, and missing values do not impede the model precision when subjected to the missing value imputation techniques.

- **Consistency**—All data fields are well-defined and managed in a consistent manner, with any changes applied to all affected databases.

- **Freshness**—Data are kept "fresh" with clearly defined rules on data shelf life. All data fields and analytics-derived results are time-stamped and periodically tested for validity. This is especially important because the mobility of today's customers and their relatively shorter careers means that most consumer data can become dated pretty quickly. According to an *MIT Sloan Management Review* article[1] titled "Does data have a shelf-life?," the data shelf life must be defined and periodic data refresh must be adopted as a part of the data governance policy. As data becomes stale, its effectiveness and capability to convey useful knowledge diminishes. Such knowledge loss in general can be estimated by the decrease in model lift.

- **Timeliness**—Data is collected, extracted, transformed, and loaded (ETL) and is available when needed.

- **Clarity and relatedness**—Data fields and their relatedness are clearly defined to permit flexibility in aggregation and combination for analytics.

Data Security and Data Privacy

As the demands of customer data by business grow, the responsibility for business to be a trusted data custodian also increases. Unfortunately, to date, much of that trust has been significantly eroded with the security breaches and privacy issues consumers encounter almost every day (as seen on the news). To ensure business has the raw ingredients (that is, data to generate tangible values for not just their own

---

1 http://sloanreview.mit.edu/article/does-data-have-a-shelf-life

business but for the consumers), the issues of data security and privacy must always be front and center of all business considerations.

Data Security

With the recent spate of data breaches to major U.S. retailers,[2] privacy and data security have become an increasingly important issue for businesses. In fact, the leading cybersecurity expert, iSIGHT Partners, found that the security breach at Neiman Marcus and Target was not isolated and likely to affect more businesses. Even as this book was going to print, the largest data breach happened at JP Morgan Chase that affected 76 million households! Not only was the breach at Target serious, but it was also sophisticated. It did not attack the servers; it attacked the point-of-sale (POS) systems. Data security needs to catch up and must cover the entire value chain of customer data from the moment a customer enters a store (online or offline) until he browses, purchases, and is offered a coupon for the next visit.

The issue is quickly evolving and beyond the scope of this book, but suffice it to say that no analytical databases should contain any personal identifiable information (PII) fields that include names, email addresses, home addresses, phone numbers, social media accounts, and logins. They can all be replaced with unique customer IDs. When needed, the other data can be linked through the master customer table, which should always be kept separate and secured from the rest of the customer data used for analytics.

---

2 http://www.dallasnews.com/business/retail/20140117-neiman-marcus-target-credit-breaches-likely-part-of-broader-hacking-attack.ece; http://www.forbes.com/sites/larrymagid/2014/10/02/jp-morgan-chase-warns-customers-about-massive-data-breach/

Data Privacy

Given the explosion of mobile and sensor data, Forrester Research, in a December 2013 report on privacy,[3] states that instead of an issue to be avoided, privacy is "essential for building trust, the foundational currency of social, mobile, and local services." Because of the ambiguity surrounding privacy and the ever-changing social norms caused by the privacy snafu of the week, most businesses adopt a "do first and ask for forgiveness later" attitude.

As a result, some companies tend to collect more data than they need, contend that "voluntary" data stewardship fails anyway, hide behind vaguely worded permissions (in most mobile apps), and adopt hard-to-understand and shifting privacy policy statements. Given the outcry over the recent companies' sharing of customer data with the National Security Agency (NSA), public tolerance of privacy infractions is getting smaller.

Because data is absolutely necessary to fuel the new economy, any change in data availability significantly hurts the economy and ultimately the consumers. A balance between the two must be attained. In fact, various legislative safeguards and consumer acceptance of the sharing of personal financial data for credit scoring illustrate just such a balance between privacy protection and the public good created through data sharing.

Contextual Privacy

Unlike data privacy safeguards on personal finance data, there are greater complexities surrounding consumer personal data in general. The complexities arise from the sheer volume, types, and number of players in generating, maintaining, and using data. Facing such complexities, I believe, it is virtually impossible to resolve it by legislation alone.

---

3 http://www.forrester.com/The+New+Privacy+Its+All+About+Context/fulltext/-/E-RES108161?intcmp=mkt:blg:ml:Prvcy_Cntxt_Rprt

A new way of thinking is therefore needed. One example is the concept of "contextual privacy." The same Forrester Research report defines contextual privacy as "a business practice in which the collection and use of personal data is consensual, within a mutually agreed-upon context, for a mutually agreed-upon purpose."

Value Exchange

To obtain consumers' consent, businesses must provide commensurate values in return. The current one-sided situation is clearly not tenable in the long run, where business reaps the lion's share of the benefits from consumer data sharing, with little or no value returning to the data owners.

How much value does business need to give up to consumers to be permitted access to their data? The Forrester Research report found that the three greatest incentives are cash rewards, loyalty points, and exclusive deals. A large portion (42 percent) of the adult consumers in the United States would never share their personal information. It is clear that unless there is a privacy resolution soon, this non-sharing consumer sector will likely grow. Once the non-sharers are in the majority, they may demand a blunt political solution to be imposed on businesses. To avoid this undesirable outcome, it is important for businesses to embrace and support privacy resolution as front and center of what they do and with some urgency.

Privacy Contexts

The Forrester Research report mentioned five contexts that businesses should respect if they want to adopt better privacy practices. Based on my experience, I would add and combine them into four distinct contexts:

- **Time context**—All data collected and used must have a specific time limit agreed upon by the consumer. No sharing should be viewed as perpetual.

- **Location**—With the prevalence of location marketing, the locations where consumers shop and where they may want to receive offers or services (or not) must be specified, agreed on, and respected.

- **Use context** —In simple terms, businesses must use the data for the stated purpose only. As an example, consumers' email addresses for operational and product-related issues should not be used for marketing purposes.

- **Extent context**—With the emergence of the sharing culture, consumers might want to share different levels of their data, depending on the extent of their social contacts. For example, a customer may tweet about a vacation destination, but share the detailed itinerary only with someone who is helping to plan the trip.

Ensuring full compliance in contextual privacy can be quite cumbersome and expensive under today's technology. However, to allay consumers' fear of losing their privacy, I propose a privacy hierarchy be defined, instead of a single privacy.

Definition of Privacy Hierarchy

I think that today's privacy definition is too broad and vague, so it has contributed to the sharing phobia of many consumers. Looking deeper into the privacy concerns, I see privacy separated into three distinct types: Access privacy, Identity privacy, and what I call "Pooling" privacy. Let me define them as follows:

- **Access privacy**—Consumers have the right to control who can contact them (granting and rescinding permission at any time) with what communication channel, when, and where.

- **Identity privacy**—Consumers have the right to not disclose their identities to certain business entities, regardless of purpose of use. PII should be defined as any data fields—by themselves or in aggregate—that reveal consumers' identities. According to the National Institute of Standards and Technology (NIST) guide to protecting the confidentiality of PII,[4] it should include "(1) any information that can be used to distinguish or trace an individual's identity, such as name, social security number, date and place of birth, mother's maiden name, or biometric records; and (2) any other information that is linked or linkable to an individual, such as medical, educational, financial, and employment information."

- **Pooling privacy**—Consumers have the right to withhold any of their data from being "pooled" together with other consumers' data for analytics. The analytics for pooling must be strictly for the purpose of finding patterns, values, and preferences of groups of consumers *without knowing their identities*. The analytics insights allow the businesses to match the right products and services to the right consumer at the right time, place, price, and cadence. The cost savings and value created are then shared with the consumers who allow data pooling for analytics.

As shown in Figure 3-1, the intrusiveness ranges from the lower level of pooling privacy, in which it is minimal, to the highest level of access privacy. By definition, the higher hierarchy implies the granting of the lower level of privacy permission. Today's key consumer privacy concern is the harm that may result from data sharing and the loss of privacy. Because there is no clear definition of privacy for consumers, any loss of privacy implies the loss of all three kinds of privacy, which inevitably conjures up the worst-case scenario in the consumers' mind: people will know who you are (identity privacy), know everything about you (pooling privacy), and be able to reach you

4 http://csrc.nist.gov/publications/nistpubs/800-122/sp800-122.pdf

by any means (access privacy). However, this would not occur if we could distinguish privacy by its three subcomponents and treat them separately. I believe that this provides a way for a workable privacy solution to be found and will be acceptable to the consumers while it will also benefit them.

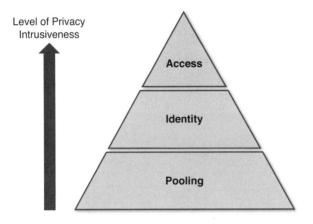

Figure 3-1   Privacy hierarchy and level of intrusiveness

A Privacy Solution

Given the definition of pooling privacy, I believe that most consumers would allow their data to be shared, provided that proper safeguards and incentives are in place. These safeguards should include the following: 1) no PII can be released, and no data can be used to trace back to the consumer; 2) any insights found must not violate any state or federal laws (such as gender, race, and age discrimination); and 3) the values created should be shared with those consumers who share their data.

To ensure that the first safeguard is properly administered and results in public trust, a technology and process solution must be devised to ensure that the collection and combination of personal data does not convey any PII. One way I suggest is to have an entity similar to the Internet Corporation for Assigned Names and Numbers

(ICANN). Instead of regulating IP addresses, this new entity would regulate the issuance of universal consumer IDs.

Consumers could grant additional permissions to reveal their identities and/or allow access. Provided that the analytics exercises are done without any PII, a business should be able to append the analytics-derived customer profiles and scores and engage their own customers accordingly. Once this is done, I see little need for businesses to know the details of any individual customer data.

Data Governance

Data governance is gaining more attention today as more and more businesses start to recognize data as a critical enterprise asset. Good data governance must cover the following disciplines: data quality, data management, data policies and strategies, business process management, and risk management. In short, the Data Governance Institute (DGI)[5] defines data governance as "the exercise of decision-making and authority for data-related matters." An enterprise data governance is necessary because many issues surrounding data cannot be left to individual managers; decisions have to be made beforehand about how to decide on matters relating to data at all levels of the company.

According to a Q1 2013 Forrester survey,[6] the importance of data governance is in supporting innovation and creating business success. Though most surveyed saw regulatory compliance as the number one driver for data management, those whose data management kept pace with business are "twice more likely to take advantage of new opportunities and create revenue streams from data."

Based on my experience, the important steps of data governance may be summarized as follows:

---

5 http://datagovernance.com/dgi_framework.pdf
6 http://www.forrester.com/Data+Governance+Equals+Business+Opportunity+No+Really/fulltext/-/E-RES83342

1. **Set up a data governance council**—Ensure that all the important concerns from every aspect of the business are properly incorporated into the official data governance documents:

   - Important parts of the business must be represented in the data governance council. Members of the council must include key leaders from IT, marketing, sales, analytics, and legal functions.

   - The council should be chaired by someone who has the ability to build consensus across the political landscape, has direct access to the CEO, and is also entrusted to be responsible for the council's progress.

2. **Evaluate the situation**—It is important to benchmark the existing situation in terms of current practices and then assess potential areas of risks, opportunities, and improvements needed. A strategic roadmap should be set up by the council to take the business to where it should be.

3. **Develop a data governance strategy**—According to the DGI Governance Framework,[7] there are ten steps for developing an effective strategy:

   - **Define the mission**—Why is data governance needed?

   - **Establish focus areas**—Determine goals, metrics, success measures, and secure funding.

   - **Formalize data rules and definitions**—To ensure clarity and alignment.

   - **Agree on decision rights**—Agree on who, when, and what process will make the data-related decisions.

   - **Establish accountabilities**—Ensure who should do what and when; document the findings to prove compliance.

---

7 http://datagovernance.com/dgi_framework.pdf

- **Ensure controls**—To mitigate risks by different levels of control according to the degree of exposure and levels of responsibility.

- **Identify stakeholders**—Identify who may affect or be affected by the data governance decisions and to solicit and address their expectations and concerns.

- **Set up a data governance office**—The data governance office (DGO) facilitates and supports the data governance council's activities.

- **Appoint data stewards**—Empanel key stakeholders to work on policy and recommend changes covering specific issues or needs.

- **Formulate a data governance process**—Establish standardized, documented, and ongoing processes to support compliance requirements for data management, privacy, access management, and security.

4. **Quantify the value of data**—Often, the value of data is taken for granted, so annual efforts in securing funding can be a constant battle in many businesses. However, given the importance of data in creating new innovations, revenue streams, and business opportunities, data governance should not be viewed as purely a cost center. The value of data needs to be properly assessed by business and automatically "paid for." An internal marketplace can be set up in which all business units "pay" to ensure that they will have the data in the right condition and customized state to support their business line.

5. **Set up a data governance scorecard**—Data governance is like dental hygiene: you don't realize you are doing a bad job until you run into trouble. It is therefore prudent for businesses to establish a consistent method of measuring progress and value and documenting compliance. A scorecard or dashboard

should be generated to monitor the state of data and reported regularly to senior management.

Data Preparation

One of the biggest infractions any analytics team could commit was to jump into building analytics models without being thoroughly prepared and explored the data. An appropriate investment of time in doing exploratory data analysis (EDA) has time and time again proven that it can save costly mistakes and provide clues and ideas for more effective analytics.

Following are the risks of not doing a thorough EDA before and after an analytics exercise:

- **Garbage axiom**—"Garbage in, garbage out." By not detecting errors in the data and accounting for their impacts, the analytics result may be erroneous and misleading and its impacts on business can be costly. Missing data can be properly accounted for through missing data imputation techniques, as discussed later.

- **Poor analytics**—Without knowing the interrelationship between different data fields and lacking the potential insights that EDA can provide, advanced analytic models may be poorly set up and ultimately fail to provide meaningful insights.

- **Wrong questions**—Many business questions can be answered through EDA and business intelligence (BI) exercises and should be done by embarking on advanced analytics. It is like peeling an onion: EDA and BI tools can help businesses ask the more obvious questions. Such answers might lead to more difficult questions that simple pivoting and BI visualization reports cannot answer. These questions can then be posed for advanced analytics to answer. Many modeling hours were wasted in answering either the wrong questions or providing answers to questions that are more suited to be answered by EDA or BI reports.

• **Insights oblivion**—One of the key benefits of EDA is the understanding of not only the data but also (more importantly) the customers and business. In fact, with diminishing direct contacts with customers because more customers buy online or via their smartphones, deliberate explorations of the customer data may be the only way to extract and acquire an intimate knowledge of them and business. Without this intimate knowledge, analytics modelers often miss insights that could lead to significant new business opportunities or customer value. For example, the bumps in the lift charts may indicate significant gaps in either data or deficiencies in customer relationship management (CRM) for a particular group of customers! I have witnessed similar insights smoothed by the team before the results were shared.

• **Theoretical insights**—Without the proper understanding of real-life business constraints and customer characteristics, the spectacular model lift results simply cannot be implemented or would never work in real life.

There are many methods you can use to prepare and explore the data. Because of the scope of the subject matter, I cover just the basic ones that are most important in my view. If you are interested, consult the numerous excellent references you can find by searching with the term "exploratory data analysis" on Amazon Books. The classic on exploratory data analysis by Tukey[8] in 1977 is still relevant today.

A few words of caution: profiling and EDA can reveal certain facts about how people shop, but may not indicate causal effects. For example, EDA may reveal those who bought artisan bread generally spend little on soda. However, spending little on soda would unlikely

---

8 http://www.amazon.com/Exploratory-Data-Analysis-John-Tukey/dp/0201076160/ref=sr_1_1?s=books&ie=UTF8&qid=1390191769&sr=1-1&keywords=Exploratory+Data+Analysis

be an indicator that a shopper would purchase artisan bread. They might not have the money or simply dislike soda.

The only way to establish causality is to run sophisticated analytics such as market basket analyses. Market basket or affinity analytics may be used to establish the likelihood that a shopper who has bought fresh produce will also buy artisan bread. To be sure, the positive model results must also be verified with actual controlled tests with shoppers. Causality is important to ensure that the correct lever is used (for example, the discount on bread to be offered to fruit buyers to trigger purchases of artisan bread). It is also important to ensure that the incremental returns are sufficient to warrant the costs of deploying the lever. In general, to investigate causal effects and predict future behaviors, you have to use advanced analytics. EDA and BI analysis reveal correlations, but do not indicate causality.

Data Manipulation

This section introduces topics on types of variables, treatment of missing values, normalization or standardization of variables, and use of data partitions. Even though this section is relatively brief compared with other sections in this book, the efforts of cleansing and preparing the data may constitute 30 to 40 percent of the analytics projects. The actual percentage would depend on the size, state, and complexity of the actual data. For more details, refer to the Appendix.

Types of Data

For analytics, data needs to be pre-specified and read in according to its type. Some algorithms try to guess the type, but it usually takes a long time and can sometimes be wrong. So it is always better to pre-specify the variable types if possible. One simple way is to create a dummy line of data at the beginning of the file with the correct formats, especially if the data contains missing values. It is important to ensure that the data is the right format so it can be read correctly.

There are several common variable types in business data:

- **Integer**—Whole numbers without decimals (such as 50 is the number of states in the United States).

- **Numerical**—Numbers with any number of decimal places (such as transactions of $1,000.50 or a propensity of purchase of 0.8145 or 81.45 percent).

- **String**—Non-numerical text data that can be names, dates, and words. String data is usually stored as simple ASCII text.

- **Categorical**—Texts that have only a few set parameters (for example, Male, Female, and Gender Neutral are the values for a category variable called Gender).

- **Nominal**—Texts that denote names such as First and Last Name, State of Residence, and so on.

- **Ordinal**—Variables in which the text values can be ranked or ordered. For example, the variable Education may be number of years of school, which is integer; or divided into Primary, Secondary, Colleges and Universities, and Post-Graduate Degrees, which is clearly ordinal (Secondary is more educated than Primary, and Colleges and Universities is more educated than Secondary). These values may sometimes be replaced with numbers (for example, 1 to 4). However, ordinal values do not indicate magnitudes (someone with a categorical value of 4 and a master's degree cannot be assumed to be 4 times more educated than someone who has only a primary school education).

Categorize Numerical Variables

Instead of dealing with numerical variables with an infinite number of possible values, it may be preferable to divide the variable into bins or categories. For example, if actual household incomes may be hard to get and are subject to large uncertainties, it may be better to bin the customers into High Income, Mid-Income, and Low Income (or any number of income bins). One added advantage is the higher likelihood for the model to differentiate behaviors between customers from different categories than between those with different incomes (for example, between $120K and $131K).

Dummy Variables

Another way to deal with nominal variables is to create so-called dummy variables in which the nominal or category variables are transformed into dummy numerical variables of 1 or 0. For example, the Gender variable, when used together with other numerical variables such as age or number of years in school, may be transformed to two variables: Male and Female. A male person would be 1 in Male and 0 in Female. By default, if both variables contain 0, the field is not specified or missing, denoting either Gender Neutral or no data.

Missing Values

Missing values should not always be taken to be zero. For example, online shoppers who browsed but did not buy or abandoned their shopping carts are not the same as those shoppers who did not shop during the same period. Depending on the context surrounding the missing values, the appropriate missing value imputation method should then be used:

- **Do Nothing**—If the missing data can safely be ignored without causing any confusion by the nodes used in the workflow.
- **Remove Row**—The missing data indicates that the other field values in the same row are of questionable validity, so the entire row is removed.

- **Min**—The default or baseline value of the entire data set for that field is taken to be the minimum value.

- **Mean**—The missing data is taken to be the average value for the particular variable.

- **Max**—The default or baseline value of the entire data set for that field is taken to be the maximum value and replaces all missing values.

- **Most Frequent**—The missing data is taken to be the same value as the most frequently occurring values for the particular variable.

- **Fix Value**—The missing values correspond to a certain value; for example, missing discount values for coupons given out during in-store tasting promotions should be taken to be the same as nonredemption, and hence zero.

Data Normalization

In many models, the ranges of the Independent Variables (IVs) can vary greatly. Some algorithms require the variables to be of similar range (in clustering, for example). In all clustering, a distance measure needs to be defined. Without standardization, a small percentage change in a variable with a large range will have inordinately larger impacts on the clustering results than the variables with smaller ranges that may be undesirable when assessing performance. There are three methods of normalization:

- By their absolute range (that is, maximum-minimum)

- By their standard deviation (that is, Z-score normalization, where Z-score is essentially the number of standard deviations the IV is from the mean)

- By their percentile ranking

Data Partitions

As part of the supervised modeling process, it is common to divide the data set into two or three samples known as partitions. The first partition is usually a Training partition (that is, used to train the model), and the second partition is usually known as a Validation partition. The validation sample is used to validate the trained model without further model adjustments.

Sometimes the second partition is used to choose the best combination of model parameters, such as the number of clusters (to prevent overfitting) or the best model out of several models. In this case, a third partition known as the Test partition is then used to test the best model chosen from the first two partitions without further changes.

- **Sampling**—The first partition can be sampled in several ways:
  - **Taking from the top**—It samples from the first row until the specified sample size is reached.
  - **Linear sampling**—Also known as the Nth sampling; a 1 percent sample is taken from the top, consisting of the set of rows {1st, 101st, 201st}.
  - **Draw randomly**—It draws a sample randomly in two ways. It can draw it without a fixed seed, which means that every time you invoke this node, it draws a different sample. However, if you want a specific randomly drawn sample that you can return to later to check the results, specify a specific seed (choose any integer number that you can refer to).
  - **Stratified sample**—A random sample is drawn while keeping the same distribution of values in a particular nominal field (for example, Gender) as the entire population. If there are no nominal variables, no stratified sampling can be done.

Hold-Out Partitions

There are two ways to train predictive models. One is to draw random samples from data of the same time period; the other is when longitudinal data that spans over a sufficiently long period of time is available for model training. Samples from earlier time periods are used to train the model. The trained model is then used to test the predictions against untouched or hold-out data in a later period. The former is usually referred to as out-of-sample training and validation; the latter is known as out-of-time model training and validation.

Here's an analogy. Suppose that you have a time machine in which you can travel across time at will. Suppose you travel back in time to 2012 and train a predictive model based on what data you have in 2012 (making sure that the data is properly quarantined for time). The predictive model trained on 2012 data is then found capable of accurately predicting 2013 results. As far as the model is concerned (assuming all things remain unchanged), the same should hold true for the model built on 2013 data and applied to predict outcomes in 2014. The model would not know the difference between predicting 2013 or predicting 2014 because both are dates in the future. The model is based on the prior year's data. The beauty of time-dependent holdouts is you do not need a time machine to test the validity of models in predicting the future. All you need is to build a series of predictive models based on past years and simply progress them one year at a time. If your models are capable of predicting the "future," then unless something drastically different happens in the coming year, there is no reason the model built on the current year's data would not be able to predict the results for the coming year!

Sometimes, a model built several months before can still be predictive by simply updating the IVs with more recent values and checking that the same model lifts remain. This kind of model validity over time is sometimes referred to as model freshness or model shelf life. As time passes, however, the underlying conditions probably change materially from the training period. As conditions change, a model's predictive power in general decays over time. It is important to monitor the lift of models over time. Depending on business requirements and how costly it is to refresh the data and rebuild the models, predictive models should be regularly refreshed with more recent data or be totally rebuilt.

Exploratory Data Analysis

Instead of exploring a multidimensional Big Data problem, let's simplify it into something we can easily handle and visualize: a three-dimensional data cube (see Figure 3-2). A larger data set simply means a cube in more than three dimensions and a much larger number of subcubes.

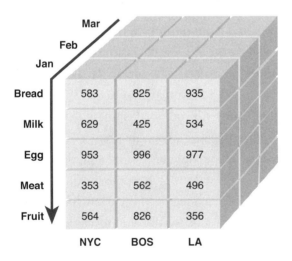

Figure 3-2   Three-dimensional data cube

Multidimensional Cube

The concepts in the next section are illustrated by using the same example. Let's assume that you are interested in the sales of your stores. The store sales in US$ are arranged by respective products (Bread, Milk, Eggs, Meat, and Fruit) and by their locations (NYC, Boston, and LA) for the months of JAN, FEB, and MAR. For starters, you could perform some simple statistical characterization of the data (for example, Min, Max, Mean, Std. deviation, Variance, and Overall sum).

In addition to the simple statistics, you may want to see how sales in each store can vary over time or by products. To answer these and other related questions, you may need to use operations such as slicing, dicing, drilling down/up, and pivoting.

Slicing

As shown in Figure 3-2, you can perform a vertical or a horizontal slice to answer different questions. To get all the sales in January or for the NYC store, you take a vertical slice along one of the two dimensions. To get the former, the vertical slice is along the month=JAN dimension.

A horizontal slice answers the question regarding products. For example, to find out how much Meat was sold, you can sum over the sales from the fourth horizontal slice from the top.

Dicing

Dicing isolates a particular combination of parameters. For example, the sales of Milk and Eggs in the Northeast region for Q1, excluding the January month (an atypical number because of unusual weather?), results in a cube of 2x2x2.

Drilling Down/Up

Assume that there are subcategories below each dimension. For example, below Month, you may get Weeks; below Cities, the Store no.; and below Meat, Meat types. You can then drill down to the subcategories, or drill up by summing the subcategories into higher categories (for example, summing Months into Quarters).

Pivoting

Pivoting allows you to answer more complex questions than slicing and dicing does. For example, if you want to know how the product sales vary for each of the three cities, and you do not care how they vary over time, you could pivot by cities and sum over the sales for the city over the three months. Instead of summing, you may want to get the average monthly sales or largest sales. More details on how to do this are discussed in the Appendix.

Visualization of Data Patterns and Trends

To process large quantities of data with hundreds or even thousands of dimensions and billions of rows, the previous methods may not be sufficient to mine insights from the data. Regardless of what specific questions you may want answers to, the questions can be answered by analyzing the baseline conditions, patterns (how the data varies and regions of large changes), outliers, or trends over time. Before more advanced analytics are discussed in the next chapters, we will explore some BI techniques, including visualization of Big Data and dashboards.

Popularity of BI Visualization

The usefulness of BI visualization can be compared with the way the automotive dashboard allows drivers to drive without knowing much about the inner complexities of cars. A good BI visualization tool must do the same for business users to drive business decision-making. As was pointed out by Atin Kulkarni, senior director of strategy and analytics at Match.com,[9] the aim for the company to start using Tableau for BI visualization is to put it "in the hands of our users, not elite analytics or BI experts." Putting vital information from different teams across the business silos on the same dashboard can help save costs of duplication of efforts by greatly improved communication.

Selecting a BI Visualization Tool

Among many criteria in selecting a BI visualization tool, I highlight the following:

- **Data access and integration**—It should be able to access data residing on different databases and provide easy tools to integrate it for analysis.
- **Charting options**—Given the complexities and multidimensional nature of data, simple pie, bar, and line charts are clearly not enough. In the very least, a visualization tool must contain the following basic charting options:
  - **Text tables**—These tables show crosstab values of two dimensions (for example, the sales price between products and cities in the previous example). Text tables may be highlighted to show color gradients (for example, the variation of margins across products and cities, from red to blue for losing money to highest margin, respectively).
  - **Scatter and line charts**—Scatter charts show data points on two axes. If the trends or trendlines are helpful, a line chart can be used. Line charts are also useful for time series analyses.

---

9 http://www.informationweek.com/software/information-management/how-to-choose-advanced-data-visualization-tools/d/d-id/1105480?

- **Box charts**—Used when the distribution (median and quartiles) of data at each point is important. The box is drawn at the first and third quartiles. The so-called whiskers are drawn at the minimum and the maximum values, but they should not be more than 1.5(Q3–Q1), lower than Q1 or 1.5(Q3–Q1), or higher than Q3, respectively. If so, the whiskers will be drawn at Q1–1.5(Q3–Q1) or Q3+1.5(Q3–Q1). Points outside the whiskers are outliers.

- **Bar charts**—Usually used to compare values across dimensions. For example, you can use side-by-side bar charts to illustrate how milk sales vary across the three cities in each of the three months. You can also add the sales targets as lines on the respective bars to indicate whether they exceeded or missed the targets (this is sometimes known as a bullet graph). You can also add 80 percent and 120 percent of the targets as gray bars against the actual sales, for example. This is helpful to see whether the sales of the particular city underperformed or overperformed with respect to the target.

- **Geographical maps**—Capability to plot all the other charts on top of a geographical map.

- **Heat maps**—Using color and size, heat maps can compare two variables.

- **Tree maps**—Used when the parameters are too many to show for bar charts (more than ten) and when some of the underlying data may need to be displayed for clarity.

- **Histograms**—Used to show how a continuous variable may be binned into discrete values.

- **Gantt charts**—Used to denote the time duration for an activity. The length corresponds to the time duration.

- **Multichart displays**—The power of data visualization often lies in the ability to analyze data from different perspectives simultaneously to detect relationships, see causal factors, and enable a holistic view of the whole data set. Once well understood, the multichart display can be constructed as a dashboard that allows the viewer to vary the views interactively.

- **Ease of use and speed**—To ensure that the tool will be adopted widely, it must be easy to learn and intuitive in its functionality. To encourage frequent use, the tool must also complete the tasks in seconds rather than minutes.

There are many BI data visualization platforms available today. Forrester Research did an evaluation of the "Advanced Data Visualization (ADV) Platforms"[10] for Q3 2012. In addition to the previous criteria, there are some other useful criteria you might need, depending on business requirements and data complexities:

- **Dynamic data content**—The ability to link to databases that are constantly being updated.

- **Visual querying**—The ability to manipulate the charts to drill down for more details.

- **Linked multi-dimensional views**—When one view is changed, the other charts on the same panel also get updated.

Among the BI vendors, Tableau Software and Tibco Software are usually regarded as the top two leaders in terms of the strength of their offerings. There are many other vendors[11] that offer cloud-based, mobile-enabled, and special-purpose solutions (for example, for research survey visualization). It is beyond the scope of this book to go into the details of how the vendors differ in their offerings, so you should check the data visualization tools against the preceding criteria and your specific requirements.

---

10 http://www.forrester.com/The+Forrester+Wave+Advanced+Data+Visualization +ADV+Platforms+Q3+2012/quickscan/-/E-RES71903?highlightTerm=data visualization platforms&isTurnHighlighting=false

11 http://www.greenbookblog.org/2013/08/04/50-new-tools-democratizing-data-analysis-visualization/

Beyond BI Visualizations

Before ending this chapter, I want to remind you that BA is a continuous quest to peel back the layers of complexities surrounding business data. BI visualization tools can answer some of the questions, but you are likely to discover many more layers below.

For example, once you see that the customers' buying behavior differs significantly from store to store, you might want to know who the customers are and why they are shopping at the stores, what their purchase histories are, and what you did or did not do to cause the trajectories of these different customer experiences and behaviors. As you can see, the list of questions can go on and on.

In peeling the onion, the innermost layers are often beyond the reach of current BI tools. They can be tackled only by using advanced predictive analytics. Some BI tools try to add analytics to their solutions, but have simply made them unwieldy and even harder to use. Because BA is still in its early days, many novel recipes are still being experimented with in different business contexts. If you do not know what to cook, it would be impossible and premature to design a Swiss Army knife kitchen gadget to replace all the kitchen tools. It is far better to focus on how to cook using the most appropriate tools we have right now. Hopefully, this is what this book can help you to do.

Conclusion

This chapter showed that before data can be fully analyzed and leveraged, businesses must do the following:

- **Integrate data**—Determine how the diverse data can be linked and integrated, either directly or inferred.
- **Ensure data quality**—Understand data quality issues and know how to discern and improve data quality for analytics.

Don't let poor data quality stop you from doing analysis. Learn to make lemonade out of lemons and become better able to quantify the value of good quality data.

- **Comply with contextual data privacy and hierarchy**—Understand the importance of data security and privacy issues. Businesses should consider the new concept of contextual privacy, leverage such knowledge, and be able to develop a sustainable privacy policy. Consistent with the policy, businesses will continue to please consumers by rewarding them commensurately, respecting their rights, and maintaining transparency on what is done with their data. Businesses should also consider the privacy hierarchy and study how separating the different privacies by context generates a win-win solution to both business and consumers.

- **Adopt robust data governance strategy**—Businesses should know how to set up a progressive and robust data governance strategy that enables the enterprise to continue to reap benefits from data and analytics.

- **Thoroughly prepare data**—Businesses should try various way of preparing data by rationalizing the different variable types, dealing with missing values, creating dummy variables, normalizing data, and partitioning data into samples for subsequent exploratory and modeling exercises.

- **Explore data**—Before conducting modeling, businesses should first analyze the data by slicing, dicing, and pivoting on a multidimensional data cube by using advanced BI data visualization. By adopting consistent criteria, businesses may choose the right tools and learn the various charting options for exploring conditions under diverse business needs. Many of the questions that will guide the subsequent analytics often surface during the EDA process. Even after the analytics results have been obtained, it may be helpful to visualize the model results and to augment the views from EDA. Together, they may be able to answer the questions that cannot be answered by simple data exploration.

4

Handle the Tools: Analytics Methodology and Tools

"Eating is a necessity, but cooking is an art."

—Unknown

Getting Familiar with the Tools

As more businesses adopt and start to consume outputs from analytics, it is important for the executives to know how the outputs were produced. Given the complexities surrounding most business applications, learning how to use analytics is similar to taking cooking lessons. The best cooking schools, such as Le Cordon Bleu College in Boston,[1] advertises these top four methods (among others) to learn how to cook:

- *Practice* hands-on training.
- *Train* under professional chefs.
- *Learn* in industry-equipped kitchens.
- *Focus* on perfecting your skills using commercial-grade tools and fine ingredients.

---

1 http://www.lecordonbleu-boston.com/Home

In my experience, the best way for business leaders to learn and master business analytics (BA) is the same: Learn and train by working under the masters, practice, and perfect your skills under real-life conditions. It is hard to learn cooking by just reading cookbooks.

Senior business leaders need to function as the master chefs in terms of creating new "recipes" for their respective endeavors. They must be able to pass on the skills to their teams to produce consistent products. Before any new item can appear on the menu, the master chef must first create, experiment on, and perfect the item. When this is successful, the team is then trained to follow the new recipe and the wait staffs (marketing and sales teams) learn how to promote it.

Master Chefs Who Can't Cook

Just as those who have never cooked before can never be master chefs, analytics leaders must be those who have done analytics before. Yet today's businesses are filled with analytics leaders who have never built a single predictive model. It is my belief that today's many failures and the mediocre results encountered by businesses in the applications of analytics have been caused by too many master chefs who do not know how to cook.

Having such non-analytics leaders may result in three undesirable outcomes: 1) a lack of innovations—the master chef does not know how to come up with new recipes, and neither can the team of tool experts; 2) no appreciation of analytics—the leader doesn't appreciate and trust something unfamiliar (that is, advanced analytics); and 3) blaming poor results on analytics. In many cases for which I have first-hand experience, the cause of failure was rarely due to the model not working. Ineffective models do not pass the model validations; they are rejected by their high errors and low lifts. The actual reasons for the failures are that either the wrong type of analytics was applied or the right analytics was applied, but its insights were wrongly executed or simply not followed.

It is therefore the aim of this chapter to introduce the most common tools to aspiring chefs. Like learning any other tools, the only way to gain mastery is to keep doing it. To cook, you just need to follow the instructions in the manual and practice using the workflows in an advanced and powerful analytics platform that you can download for free (http://www.KNIME.com)! A word of caution for executives: Although KNIME is free, is not a toy. It can potentially change your career path and make you, like me, into a quant or geek executive! You will also enjoy it and lead a more fulfilling career!

Types of Analytics

It is generally agreed that analytics can be grouped into four types of activities by the types of solutions provided: 1) descriptive, 2) diagnostic, 3) predictive, and 4) prescriptive. The four types can also be grouped into three stages: descriptive and diagnostic analytics are usually referred to as *analytics 1.0*, predictive analytics as *analytics 2.0*, and prescriptive analytics as *analytics 3.0*.

I furnish a brief survey of the various tools you can use to obtain the various types of solutions.

Descriptive and Diagnostic Tools: BI Visualization and Reporting

In a report (2/20/2014)[2] on the "Magic Quadrant for Business Intelligence (BI) and Analytics Platforms," Gartner divided BI from advanced analytics[3] for the first time because the two have grown to

---

2 http://www.gartner.com/technology/reprints.do%3Fid=1-1QLGACN%26ct=140210%26st=sb

3 http://www.gartner.com/technology/reprints.do%3Fid=1-1QXWE6S%26ct=140219%26st=sb%23

serve different customers, and the criteria of functionality for BI and advanced analytics have become quite different.

For example, BI and visualization tools were mainly designed for business users. How well these BI tools fit your needs may be evaluated based on their reporting and dashboard capabilities; capability to analyze data; and how well they can integrate data sources, support data management, work collaboratively, and support Big Data.

Regardless of specific tools, they generally include at least some of the following features:

- **Tables**
  - **Cross (or pivot) table**—Used to visualize multidimensional data (for example, to see sales that depend on stores, date range, regions, customer segments, promotions, and margins).
  - **Graphical table**—Used to display multiple charts within the cells of a table. Although it contains rich information, it can easily appear cluttered. To avoid this, the graphics within the same table must be consistent, simple, and intuitive at a glance. You can convey trends within the elements of a table by using sparklines, which are small and simple line graphs with as few graphical elements as possible.
- **For comparisons**—When sets of results are compared, you can use one of the following graphs, depending on the following conditions:
  - **Two dimensions (for example, sales and date)**
    - For few quantities, use pie charts (for example, comparing sales of five stores).
    - For many quantities, use bar or column charts (for example, comparing sales of five regions each containing more than ten stores).
  - **More than two dimensions (for example, percent margin, revenues, and regions)**

- Use bubble charts (for example, representing the variations of percent profit margin by regions with the size of the bubble representing the size of the revenues).

- **For showing trends**—Use line charts or sparklines (to be viewed at a glance).

- **For showing the spread of data**—Use scatter charts. The trend lines with fitted curves may be added to indicate closeness of fit.

- **Maps**—For showing the distributions of results on either geographic maps or websites if the following attributes matter:

  - If actual locations are needed, display the results on a *geographic map*.

  - For showing data hierarchy, use a *tree map*.

  - For showing intensity variation, use a *heat map* that displays different intensities (for example, the portion of the website that has the most clicks or the amount of time shoppers linger across the layout of the store) in terms of color variations.

Advanced Analytics Tools: Prediction, Optimization, and Knowledge Discovery

Instead of business users, the advanced analytics tools were designed for knowledge users to perform complex analytics for knowledge discovery on multistructure data. However, one of the goals of this book is to eliminate this distinction by empowering business users as knowledge users.

A Unified View of BI Analysis, Advanced Analytics, and Visualization

Right now, business and knowledge users are probably using different tools to crunch and view the same data. To ensure sharing, both types of users could ideally analyze the same data sets on the same platform. Unfortunately, no single platform exists today that can be used collaboratively by all the different types of users. The common practice is to use the BI visualization platform for simple visual analyses and the advanced analytics platforms for performing a sequence of more complex analytics.

Although these two tasks can be performed by the same team, they are usually done by different teams. When different teams are involved, I have often seen some critical information and knowledge getting lost in the process of transferring results from one platform to another. For example, the many ugly lift charts that analysts tried hard to smooth before showing to their managers may actually contain important insights about a key customer group that were neglected and are therefore missing some critical predictors. Without seeing the results, the senior decision scientists have no way of discovering this oversight.

In a unified analytics and visualization platform, the challenges of governance, scale, and performance need to be properly addressed. Such a platform must also enhance capabilities to analyze multistructure data from diverse sources such as websites, mobile devices, sales logs, and sensors. The same platform must also be able to perform new analyses such as link/network analysis, sentiment analysis, and machine learning algorithms. More and more businesses are starting to realize new values and opportunities through such analyses.

Among the leading vendors, I discuss and use one of the four advanced analytics leaders in the magic quadrant: KNIME. An open-source tool, KNIME is used in this book to enable business users to see and experience how advanced analytics work. Let's begin the knowledge discovery journey.

Two Ways of Knowledge Discovery

There are two types of knowledge discovery processes: supervised and unsupervised. Both types are necessary and important for advanced analytics. The differences are shown in Table 4-1.

Table 4-1   Supervised vs. Unsupervised Knowledge Discovery

	Supervised Learning	Unsupervised Learning
Types of knowledge	Generally known but need to be quantified	Unknown or hidden
Targeted outcome	Dependent variable (DV)	No DV
Model training	Use samples or data partitions	No training (except for determining number of clusters)
Model validation	Validated against data holdouts (not used for model training) from different time (out of time) or sampled (out of sample) partitions	Business knowledge and usefulness of insights
Model accuracy or efficacies	Lift charts, receiver operating characteristic (ROC) curves, confusion matrix, percentage accurate prediction	Bayesian Information Criterion (BIC)
Examples	Regressions and classification	Clustering, PCA, and hidden Markov models

KNIME Advanced Analytics Platform

Not only does KNIME possess the power of a full-featured analytics tool such as the SAS Enterprise Miner, but it also offers other advantages, discussed as follows.

WYSWYG (What You See Is What You Get) Workflow

Analytics on KNIME consists of nodes connected into workflows. Specific analytics are defined within the nodes through the underlying configuration details. The workflow visually lays out the analytics process without needing any coding (unless you want to embed and hide other codes within metanodes).

This feature not only makes the analytics process apparent, but it is also useful for fostering communications and collaborations across enterprise and functional silos. For example, the data analysts could collapse their data preparation workflow nodes into a single metanode labeled as ETL node. Similarly, the data scientist could create exploratory data analysis (EDA), model calibration, model validation, and test-and-learn metanodes. Analytics strategists would then conduct their what-if analysis within the respective BI and visualization nodes within the strategic analysis metanode.

An analytically empowered business executive, who is also an analytics decider, can see the entire analytics process from this metanode workflow. The analytics executive can drill into any metanode by simply opening it, which encourages transparency and ensures that things are integrated and functioning properly. By using a common language and platform, it operationally enables the teams to connect seamlessly their work in a single workflow and be therefore jointly responsible for the eventual analytics and business outcome. It permits a process akin to the concurrent engineering[4] approach successfully pioneered by NASA and many practitioners of complex workflows.

Adaptability, Versatility, and Frequent Updates

KNIME has the capability to work with codes written in other popular languages such as R, Python, and Java through the so-called "snippet" nodes.

Its versatility is reflected in the popular Weka analytics tools and many tools developed by the KNIME community.

---

4 *Simultaneous Engineering for New Product Development: Manufacturing Applications*, J. Ribbens, (New York: Wiley, 2000).

KNIME typically updates every couple of months and constantly expands into the area of sentiment, natural language processing (NLP), and linkage with current Big Data formats (for example, Hadoop and Hive) to enable in-database analytics.

Before introducing various analytics tools that are most useful for business applications, I will show you how predictive models are built, trained, and validated.

Types of Advanced Analytics and Applications

Regardless of complexity, most business questions may be broken down into a series of simpler questions. Analytics tools can then be used to answer these questions. The list of tools discussed in this book is not meant to be exhaustive, but includes the tools that were found to be most useful. In fact, it is common in my experience to try different modeling tools to see which offers the most robust and insightful results. Given the ease of trying different models, it is rare to settle on the final result from running just a single model.

Analytics Modeling Tools by Functions

Most business questions today can be answered by a combination of models used to predict one or more of the following four outcomes:

- Likelihood of certain occurrence or events
- Grouping of people or entities
- Values of transactions; wallets; LTVs; or other measures such as number of visitors to stores or websites, or number of customers churned in the past week
- Others, such as the meaning and attributes of web contents, connectivity of social networks, sentiments, influence of a blogger on the immediate social network, and degrees of influence

Instead of focusing on the model tools, I want to focus on the business questions and how the tools may be used to obtain the solutions. (For details on how to build and run the various models, please see Chapters 6, 7, and 8, and Appendix A on KNIME.)

Modeling Likelihood

This is one of the most common applications for analytics for predicting the probability of certain occurrences and for rank ordering targeting priorities.

- **Business questions**—The propensity of any customer to do the following:
  - Respond to certain campaigns, communications, and offers.
  - Buy specific products within a certain time period, either for the first time, repurchasing within a certain period, or buying more accessories (cross-selling) or upgrades (upselling).
  - Become loyal (for example, when properly nurtured will become high-value loyal customers).
  - Churn and stop buying. Who are most likely to cancel their subscriptions or withdraw all their funds within the next month?
  - Join a particular group of targeted customers. Which customers look most like to be the early adopters of your new enhanced product?
- **Potential business applications**—The modeling of likelihood may be as follows:
  - Targeting in marketing campaigns, such as credit card offerings, customer acquisition, retention, and reactivation campaigns by targeting only those above a certain propensity to achieve the greatest returns.
  - Look-alikes of customers with a certain behavior or attitudes.

- Spam detection through training an artificial neural net (ANN) to detect whether an email is spam.

- Formation of a posteriori causal control group by the technique known as *Propensity Score Matching* (PSM). This is a powerful technique invented by a couple of statistics professors from Harvard and University of Pennsylvania in 1983[5] to de-bias two samples for comparison by the use of a single measure (that is, the propensity scores of belonging to the target group). PSM is often used to construct control groups after the fact when holdouts are impossible or too expensive. This technique was widely used in the banking and financial sectors due to the high costs of control groups held out from beneficial treatments. There are conditions that need to be met before the biases can be fully accounted for or real incremental effects isolated. If possible, a multicell design of experiments (DOE) should be used to establish causality and incrementality.

- **Analytics Model**—Logistic regression or ANN (used when a black-box solution is acceptable).

- **Dependent variables (DVs)**—Probability of the customer being the target.

- **Independent variables (IVs)**—Examples of possible IVs are commonly available characteristics in the customers' demographics (age, gender, education, marital status, children), what they purchased before, online browsing history, and so on. Additional data on hobbies, life styles, household assets, and incomes may be purchased from data vendors such as Experian or Acxiom.

---

5 Rosenbaum, Paul R. and Rubin, Donald B. (1983). "The Central Role of the Propensity Score in Observational Studies for Causal Effects." Biometrika 70 (1): 41–55. http://biomet.oxfordjournals.org/content /70/1/41.

- **Training sample**—The model is usually trained by predicting accurately which are the targeted customers from a sample of targets (reference category [for example, churned = 1]) and nontargets (churned = 0). The training sample may be a partition from a larger sample (out of sample) or from a previous time period (out of time).

- **Validation on the holdouts and lift charts**—The trained model is then validated on the sample held out from training the model (therefore, a holdout sample that can be out of the same larger sample or a sample later in time than the training sample).

A *lift chart* is a convenient way of showing how a predictive model performs. The lift chart shown in Figure 4-1 is for a model to predict who the females are among the targets. The business context may be based on some observed online behaviors (IVs); a model was built to predict which visitors should receive offers for female products.

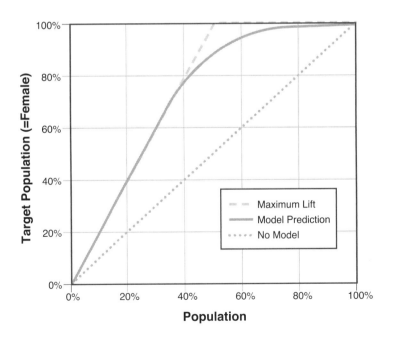

Figure 4-1   Model lift chart

Based on their respective IVs, visitors are given probabilities of being a female. A lift chart is then constructed by first ranking the visitors in descending order of their probability of being a female. The percentage of those visitors accurately predicted to be female is then plotted on the vertical axis. Assuming that tossing a coin is used as the model, the primitive model would find x percent of all the female visitors in the sample for x percent of visitors. The blue line indicates the results of the primitive random model.

If the analytics model is any good, it must produce a lift when compared with the primitive random model. In this case, the lift is shown as the red curve and is clearly very good. The lift is almost constant at 200 percent, meaning that by serving only 20 percent of the visitors, you can obtain 40 percent of all the female visitors. The visitors are assumed to consist of equal numbers of females and males. After serving 50 percent of the total visitors, you can be pretty confident that at a 200 percent lift, the model has captured every female single visitor in the population. So the yellow line, in fact, is the performance of the ideal model or the perfect crystal ball model. No model applied to this case can perform better than a 200 percent lift.

You need to be cautious when comparing lifts of different samples. Even in a perfect scenario, the model can have a maximum lift of only 200 percent. In contrast, for online or retail buyer targeting, the conversion is usually 0.1 percent or less, so the maximum lift can be as high as 100 percent/0.1 percent, or 1,000 or 100,000 percent! A lift of 30 percent may sound like a large lift, but it is actually quite miniscule, and the lift curve barely lifts off from the blue baseline.

As more model performances are expressed in lift charts, there are a few useful tips to prevent misleading:

- **Comparing training and validation lift charts**—A lift chart should not be viewed in isolation. The lift chart of the training sample must be compared with that of the validation sample. If the two lift charts show very different shapes and magnitudes, the model is highly dependent on the particular samples chosen, so it is likely not predictive.

- **Missing major IVs**—In an effort to obtain a good lift chart with a smooth and large lift, lift charts that contain humps at higher deciles are often discarded as erroneous. However, this kind of chart may contain useful information. For example, it may actually be indicative of the model missing major IVs. The humps may simply be from those targeted customers that cannot be modeled using the existing IVs. The unaccounted-for targets within the humps should be isolated, profiled, and analyzed to see whether it is indeed the case.

- **Cross-validation**—Although a nice thing, a single smooth lift chart with a large lift may simply be an isolated instance. When the model is deployed in real life, it then produces limited or even zero lift. A safer approach is to do a cross-validation in which many different models are trained and validated using different randomly selected samples. The overall model performance can then be reliably inferred as the average performance of the models.

Modeling Groupings

In many applications, grouping people or entities to detect group characteristics is usually done first. For example, before you can answer any detailed questions about individual customers, you need to view them in groups. The groups not only simplify the analysis, but you can also see distinct behaviors and attitudes more clearly in groups than with individuals. My analogy is that it is easier to find a necklace of diamonds in a beach than to find loose diamonds within pebbles and sand.

Once insights are found for the groups, the insights can be prioritized by the relative sizes and values of the groups, which makes the group or segment insights much more actionable. If the customers' characteristics group themselves, the groups tend to exhibit similar responses and personas when touched and when interacting with the brand and products.

There are two ways to group the customers. In supervised learning, you can understand the differences from those who responded to the recent promotions versus those who did not. As its name suggests, in unsupervised learning, the grouping is done without any prior notion of any outcome. Instead, customers are grouped based on how similar they are according to the segmentation variables.

Supervised Learning

Supervised Learning is used when businesses need to group customers or entities based on some measured and known predictors (IVs) to predict a certain unknown outcome (DVs).

- **Business questions and applications**—What are the differences between groups of customers and the rules for predicting which group any customer is likely to belong? Some examples include the following:
  - Responders and nonresponders to certain campaigns, communications, and offers.
  - Buyers and nonbuyers of specific products within a certain time period either for the first time, repurchasing within a certain period, or buying more accessories.
  - Loyal and less-loyal customers (for example, occasional customers' profiles match the high-value loyal customers and will become high-value loyal customers when properly nurtured).
  - For churners and nonchurners: Determine the major early indicators of which customers will be most likely to cancel their subscriptions or withdraw all their funds within the next month (churners), differentiated from non-churners.

- Belonging to any particular group of targeted customers: Determine which customers look most like the early adopters of a new product.

- **Analytics Model**—Classification decision tree (C4.5 in KNIME) or any other classifier. (There are many other types of classifiers, such as the K Nearest Neighbor [KNN] classifier, or tree or random tree and forest models.)

  - **Dependent variable (DV)**—Text labels indicating various outcomes.

  - **Independent variables (IVs)**—Examples of possible IVs are commonly available characteristics in the customers' demographics (age, gender, education, marital status, children), what they purchased before, online browsing history, and so on. Additional data on hobbies, lifestyles, household assets, and incomes may be purchased from data vendors such as Experian or Acxiom.

    Let's illustrate with a simple case of riding mower ownership (with a workflow shown in Figure 4-2) for 24 households as a function of their incomes and lot sizes.[6] (For details on applying analytics to detect patterns and rules, please see Chapter 6.)

  - **Training sample**—The model is usually trained by predicting who the target customers are in the samples (reference category or class variable in this case = owner) and nontargets (non-owners). The training sample may be a partition from the overall data set (usually smaller than the validation sample to ensure robustness. In the riding mower case, 50 percent is chosen due to the small data size).

  - **Validation on the holdouts and lift charts**—The trained model is then validated on the sample held out from training the mower ownership prediction model.

---

6 http://mineracaodedados.files.wordpress.com/2012/07/data-mining-in-excel.pdf, page 99.

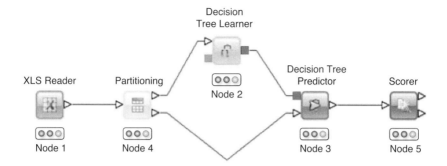

Figure 4-2 Decision tree workflow

- **Decision tree (learner or training) model**—The result of the training or learner model produces a tree, as shown in Figure 4-3 (using no pruning and a minimum number of records of two per node). One key advantage of a decision tree model over logistic regression is its capability to generate simple rules that can be easier to explain than a fitted logarithmic equation of variables and coefficients. This fact should not be overlooked because clarity is often more important than higher model lifts to business owners, especially in the earlier phase of analytics in which business owners need to get a feel for the analytics before trusting its results!

- **Decision rules**—By reading from the top, the top node shows that the 50 percent training partition (with a random seed of 1234) contains 6 owners. The most important discriminator is Lot_Size, and the split was found to be 18.8K sq. ft. The next variable is Income. Each time you move to the next level, you used the conjunction AND, so for the four terminal nodes of two owners and two non-owners, the rules for the households are the following (from the top rightmost node):

  - **Households with large lots (3/12, or 25 percent)**— Households with lot sizes larger than 18,000 sq. ft. are likely to own riding mowers.

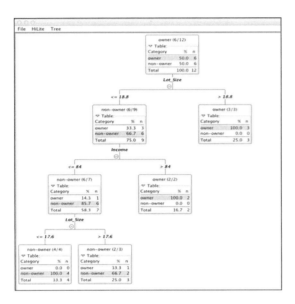

Figure 4-3   Decision tree model

- **High-income households (2/12, or 17 percent)—**
  Households with incomes greater than $84K are likely to
  own riding mowers.

- **Lower-income households (7/12, or 58 percent)—**
  The last two groups at the lowest level were classified as
  non-owners, so they can be combined into a single node.
  They are the households with lower incomes less than
  $84K.

  Even though the rules seemed to be intuitive and overly
  simple, two critical insights would be missing without analytics. Common sense would not be able to 1) place the
  exact splits in terms of lot size and income levels, and 2)
  provide the sizes and proportions of the various types of
  owners. The latter is important for the purpose of deciding the level of investments and attentions by their relative
  sizes, values, and importance to the business.

- **Accuracy of predictions—**The trained model is then fed
  into the decision tree predictor node to validate the model by

predicting the holdout sample (as shown in Figure 4-4). The accuracy was found to be 75 percent. Detailed errors may be displayed in a so-called confusion matrix (as shown in Table 4-2). Out of the 25 percent (3) wrongly classified households, they are all false positives (that is, non-owners predicted as owners).

Row ID	D Income	D Lot_Size	S Ownership	S Prediction (Owner..
Row2	64.8	21.6	owner	owner
Row5	110.1	19.2	owner	owner
Row7	82.8	22.4	owner	owner
Row9	93	20.8	owner	owner
Row10	51	22	owner	owner
Row11	81	20	owner	owner
Row12	75	19.6	non-owner	owner
Row13	52.8	20.8	non-owner	owner
Row15	43.2	20.4	non-owner	owner
Row17	49.2	17.6	non-owner	non-owner
Row18	59.4	16	non-owner	non-owner
Row20	47.4	16.4	non-owner	non-owner

Table "RidingMowers.xls [Data]" – Rows: 12

Classified Data – 3:3 – Decision Tree Predictor
File

Figure 4-4   Decision tree predictor results

- **Relative costs of errors**—In business applications, the costs of false positives and false negatives are not always the same. The general rules are these: If the costs of treatment for the targets are high relative to the values (low returns on investment [ROIs]), the false negatives are more tolerable than false positives. Whereas for low-cost touches, such as email to high-value targets (high ROIs), false positives are acceptable (that is, if you were to email a few more, you would reach more high-value targets, even if you inadvertently also emailed some who were not the high-value targets).

- **Penalty costs**—If businesses include the overall costs of spam in ruined relationships and reputations, the high percent of false positives might quickly become unacceptable. For this reason, businesses might want to specify a penalty cost to false positives to prevent the inadvertent spamming of high-value customers.

- **Leveraging false positive results**—A false positive in model results does not always indicate an error; it may be an indication that the false positives look very similar to the actual positives in terms of the predictors, but the customers did not buy under the current treatments. By further study of these false positives, you might find ways to entice them; by using better messages, higher perceived values, and enrichment offers, these false positives may become actual positive buyers!

Table 4-2   Confusion Matrix

	Predicted Owner	Predicted Non-Owner
Actual owner	6	0
Actual non-owner	3	3

Unsupervised Learning

As mentioned before, decision tree models are desirable because they can provide detailed and easily understandable rules for differentiating different classes of targets (that is, owners and non-owners). However, sometimes you might want to examine how many types of households there are without discriminating whether they have bought from you before. This is desirable because it does not include the potential biases associated with all the effects of what you have done to date. There are many options of clustering tools in KNIME (see Figure 4-5).

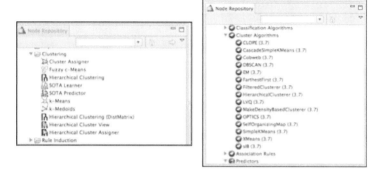

Figure 4-5   KNIME clustering nodes

For the purpose of using unsupervised learning for more effective management of customer relationships, businesses can apply to answer business questions (described in the following sections).

- **Business questions and applications**—Unsupervised Learning is used when businesses want to know what different types of customers there are and the rules for predicting to which group or segment any customer is likely to belong.

- **Analytics model**—K-means or hierarchical clustering: Hierarchical clustering tends to take a lot longer to run than k-means. There are also x-means in Weka 3.7. X-means 3.7 is used instead of k-means in this example for the following reasons: 1) it provides the model performance estimates (distortions and Bayesian Information Criterion [BIC]); 2) unlike k-means, with which you need to prespecify the number of clusters, x-means iterates between the input range of number of clusters. Based on the BIC values, x-means automatically chooses the k-means model with the number of clusters that produces the lowest BIC or distortions.

- **Clustering parameters**—These parameters include customers' demographics, online or offline behaviors, attitudes, values, and purchases. Note that if current purchase amounts are included in the segmentation, it may not be useful (for the purpose of growing wallet shares, for example, because customers with similar predicted wallets are grouped into different segments if their purchase amounts are different). A better approach is to cluster by factors that are intrinsic to the customers' underlying persona and profiles. Examples may be the customers' demographics, household composition, life events or stages (just married, pregnancy, birth, divorced, or retired), attitudes, online browsing history, and predicted wallets and LTVs.

- **Distance measures**—K-means requires a distance measure, so the distances of a particular cluster are closer to the centroid of that particular cluster than the centroids of other clusters. In the past, the distance is usually defined as a Euclidean distance (that is, the distance between 2 points (x_1, y_1) and (x_2, y_2) in a 2-dimensional space is equal to $Sqrt(Sq(x_2 - x_1) + Sq(y_2 - y_1))$. However, there are other distances that may be more appropriate for other types of variables. For example, between two nominal variables, you may want to define some similarity index. If only movements along grids are permitted, the so-called Manhattan distance can be used. Instead of the Euclidean distance, which is for a bird, the distance for a pedestrian to walk from point 1 to point 2 in NYC is $(x_2 - x_1) + (y_2 - y_1)$.

- **X-means example**—To illustrate the application, let's again use the riding mower case. Assume that you own a riding mower company. Your business has stagnated in a declining market, and you need to better understand the customers in the current riding mower market. You engaged your market research team, and you have assembled an impressive amount of customer data through a combination of customer surveys and ownership records. Before you do a complex and in-depth analysis, you decide to ask your team to give you a small representative sample of the overall data set (in this case, very small). To make things even easier to visualize, you ask that they give you only the two most important variables: Lot_Size and Income. Instead of asking your team, you decide to try a simple clustering using x-means on KNIME with just the two variables. (At this point, you should download KNIME and set up the model as shown in Figure 4-6 to see whether you can replicate the following results.)

Assume a random seed of 123 and the range of the number of clusters to be between 2 and 5. The x-means model selects a three clusters result (shown in Figure 4-7) as the best option. The centers of the clusters are as follows: Cluster 0 (Incomes = $69.19K, Lot_Size = 21,090 sq. ft.); Cluster 1 (Incomes = $96.90K, Lot_Size = 17,800 sq. ft.); Cluster 2 (Incomes = $54.87K, Lot_Size = 16,844 sq. ft.).

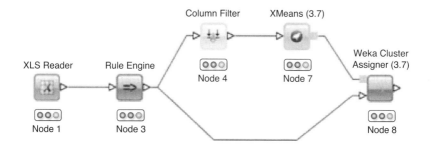

Figure 4-6 Weka clustering workflow

- **Applications**—From the clustering result, Cluster 1 is composed mainly of owners with higher incomes (except one with an income of $84K), and Cluster 2 includes non-owners with lower incomes. Despite being overly simplistic, this case actually reflects what is commonly encountered in real businesses: Some of the clusters are likely to elicit a "So what?" response from business owners. It is clearly no surprise when you report that high-income households are owners of riding mowers and low-income households are non-owners.

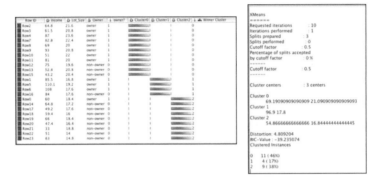

Figure 4-7   X-means clustering results

What is interesting is Cluster 0, which contains households with large lot sizes and consists of mostly riding mower owners. However, 27 percent (or 3) of Cluster 0 households are non-owners. Two are clearly low-income households and unlikely to be riding mower owners despite owning large lots. However, one household may be an acquisition target because it has an income of $75K and a lot size of 19,600 sq. ft. If you assume that the sample is representative of the real market, 9 percent of Cluster 0 or 4 percent of all households are likely owners of riding mowers but have yet to own one. Assuming a market size of $2B, 4 percent represents $80M per year of new revenues. You then decide to ask your team to refine the models with the entire data set and the other IVs to come up with a scoring model.

- **Scoring target list**—The algorithm is used to score a contact list purchased from Experian for prospects within the target segment that have been predicted to have the highest propensity to buy a riding mower this year. Multitouch campaigns including email and outbound calls are used with multicell control groups to fully test the best combination of messages, promotions, and scripts to achieve the highest conversion rates. Once optimized, the acquisition costs due to the maximized high conversion rate are reduced to less than $100. You then decide to do a full-scale rollout of the acquisition program!

Value Prediction

Businesses often need to know and predict the values of entities, actions, or customers. When prediction of values that depend on certain variables, we can use the ordinary least squares (OLS) estimator to obtain the linear equation of the DV (to be predicted) as a function of the IVs (predictors). Another technique that can sometimes give more accurate predictions is the ensemble regression tree (tree ensemble learner and predictor [regression]). Ensemble models run more than one model and then take the average of the model results. We will go over details of analytics as applied to a real case involving predicting housing prices in Boston as referenced in the following list:

- **Analytics**: Details of analytics are described in the following list:
  - **Model**—Linear regression (as shown in Figure 4-8) or ensemble regression tree model (as shown in Figure 4-9).
  - **DVs**—The values to be predicted: quantities, prices, wallets, LTVs, and any other numerical variables.
  - **IVs**—Examples include commonly available characteristics in the customers' demographics (age, gender, education, marital status, children), previous purchases, online browsing history, and so on. Additional data on hobbies, lifestyles, household assets, and incomes may be purchased from data vendors such as Experian or Acxiom.

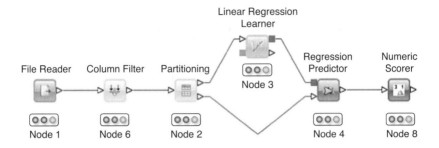

Figure 4-8   Linear regression workflow

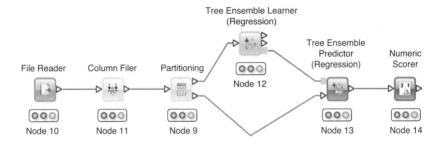

Figure 4-9   Tree ensemble workflow

Let's illustrate by using the classic case of predicting Boston housing prices in 1978.[7] (Details about the case may be found in the original paper.)[8]

- **Boston housing DV and IVs**—The DV is the median housing price (MEDV) in $1000s. To be closer to the price in Boston today, you may want to assume that the values are actually in $10K rather than $1K. The IVs are shown in Table 4-3.

Table 4-3   List of Variables

IVs	Descriptions
CRIM	Per capita crime rate by town (1970 FBI data # crimes per 1,000)
ZN	Proportion of residential land zoned for lots over 25,000 sq. ft.

7 https://archive.ics.uci.edu/ml/machine-learning-databases/housing/housing.data
8 Harrison, D. and Rubinfeld, D.L. "Hedonic Prices and the Demand for Clean Air." *Journal of Environmental Economics and Management*, Vol. 5, 81-102, 1978.

IVs	Descriptions
INDUS	Proportion of nonretail business acres per town
CHAS	Charles River dummy variable (1 if tract bounds river; 0 otherwise)
NOX	Nitric oxides concentration (parts per 10 million)
RM	Average number of rooms per dwelling
AGE	Proportion of owner-occupied units built prior to 1940
DIS	Weighted distances to five Boston employment centers
RAD	Index of accessibility to radial highways
TAX	Full-value property-tax rate per $10,000
PTRATIO	Pupil-teacher ratio by town
B	1000(Bk - 0.63)^2, where Bk is the proportion of blacks by town
LSTAT	Percent lower status of the population

- **Training sample**—The model is usually trained by accurately predicting the values of the target variable in the samples (MEDVs). The training sample is taken to a 40 percent random sample with a seed of 1234. (Again, you are encouraged to download KNIME and replicate the results.) For the ensemble regression tree, the default setting and specifying MEDV as the target column is used. Two of the IVs (LSAT and B) reflect the time the data was collected. By removing these two IVs, the model performance decreases by about 10 percent in terms of R^2. These two IVs will be removed from the models as these kinds of data are likely not available today.

- **Validation on the holdouts and model accuracies**—The trained model is then validated on the 60 percent sample held out from training the model. For the regression predictor, you need to check the box of the Prediction column to append the predicted MEDV. For the ensemble regression tree predictor, the predicted prices are in the column tree ensemble responses.

- **Model performances**

  - The numeric scorer node results as shown in Table 4-4 indicate that the ensemble regression tree model performed quite a bit better than the OLS linear regression model. By adding more models as shown in Figure 4-10, the ensemble model's accuracy increased logarithmically to about 0.76 at about 10 models.

  - One added advantage of the ensemble regression tree model is the ability to indicate which variables are more important than others, as shown in the learner tree in Figure 4-11. Variables are at the top are more important in affecting the outcomes than those below them.

  - The regression tree of Model 1 of the 50 ensemble models run (as shown in Figure 4-11) shows that the housing pricing in Boston is most affected by the number of rooms. Higher-value homes are those with more than eight rooms. For houses with fewer than seven rooms, the next factor is crime rate at 9.28 per 1,000. When crime rates are greater than 9.28, the average housing price is only half of those with lower crime rates.

Table 4-4   Model Accuracies

Model	Linear Regression	Ensemble Regression Tree (Average of 50 Random Forest Models)
R^2	0.657	0.778
RMSE	5.182	4.165

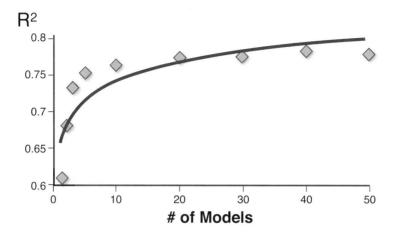

Figure 4-10   Effects of number of ensemble models

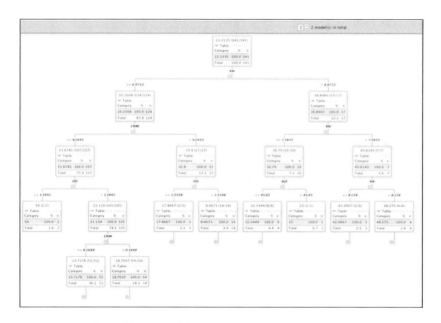

Figure 4-11   Ensemble tree model

Other Models

With the advent of new technologies and changes in consumer behaviors, many other types of data and analytics are available for the first time to businesses, including:

- **Business applications**—Analytics of new data in the following section has gained increasing business applications.
  - **Text contents**—Better understanding and tagging of website and mobile contents and autogenerating more appropriate product and site attributes.
  - **Consumer–generated text contents**—Use of blogs, product comments, evaluations, tweets, and Facebook personal and fan pages to better measure consumer interests and sentiments.
  - **Social media**—Connectivity between consumers as in their social graphs; how they interact through the new and old media to better gauge opinion, influence, and viral propagation of views and behaviors.
  - **Influencer identification and engagement**—To know who the opinion leaders are among the customers, to estimate their degrees of influence, and to devise effective monitoring systems and strategies to more proactively engage them. Increasingly, I am aware of successes of more analytics-driven marketing programs that attempt to "activate" the unpaid influencers to become advocates for the brand among their social networks. Influencer management will increasingly be the direction social media marketing will need to embrace in the future to be effective.
- **Analytics**
  - **Models**—KNIME offers many tools and example workflows for doing analytics in this area. You should download the many examples within the KNIME Explorer as shown

in Figure 4-12 (right-click Login and then drag the example workflow you want to the local or download it as a zip file).[9] You can also read many whitepapers on advanced topics on the KNIME websites and elsewhere on the Internet.

Figure 4-12 KNIME example workflows

- There are more than 60 nodes under Text Processing within the KNIME repository. Although it is beyond the scope of this book to cover every single node, I list here some simple workflows to illustrate how various tools can be used in specific business applications. With these example workflows and the whitepapers, you can follow the steps and see the way these kinds of analytics are done. It is again like learning how to cook gourmet dishes by working with the master chef. However, without overwhelming you, I will list the workflows and the results, and focus on the business implications without getting into the technical details. If you want to pursue these topics, you are encouraged to refer to the more technical references on analytics and data mining.

---

9 https://www.knime.org/knime_downloads/publicserver/publicserver_workflows.zip

- **Text mining**
  - The workflow with the data can be downloaded from http://bit.ly/Wicoyj. The zipped workflow can then be imported by using File → Import KNIME Workflow.
  - The data represents a collection of 1,000 customer reviews from Amazon Food that includes different types of food (from Stanford University).[10] The original data consists of 568,454 reviews for 3 years (October 1999 to October 2012).
  - The reviews, after excluding non-human foods and also including only bad reviews with 1 or 2 stars, were subjected to a series of text mining steps. The relative occurrences of the keywords were used to create two tag clouds: one for adjectives/adverbs and another for nouns. A part of speech (PoS) filter was used to include only texts that meet the POS criteria.
  - The data contains ProductID, UserID, Score (number of stars), text of reviews, and good reviews (3–5 stars) or bad reviews (1 or 2 stars). I briefly describe the workflow shown in Figure 4-13, and you are encouraged to load the workflow and follow along with these steps to read in the data file:
    1. Use a CSVWriter node to write the read comma-separated value (CSV) data into a new CSV data file on your local drive.
    2. Reset the data (right-click the FileReader node and then select Reset). The traffic light turns from green to red.
    3. Right-click the FileReader node and select Configure. Browse for the CSV file created on the valid URL; select "," for the column delimiter; and check the read column headers.

---

10 https://snap.stanford.edu/data/web-FineFoods.html

4. Click OK and then right-click FileReader to execute. Once done, the traffic light turns to green, indicating that the workflow is ready to move to the next step.

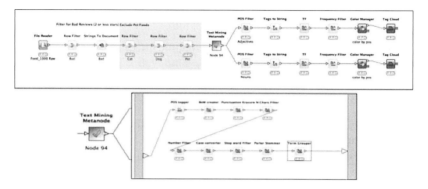

Figure 4-13 Text mining workflow on Amazon fine food review

After the data is read, I will review the steps one by one, as follows:

- The first RowFilter node is for selecting only those with bad (1 or 2 stars) reviews. (You might want to select good reviews as an exercise.)

- The StringToDocument node changes the CSV table into a document table readable by the subsequent text mining nodes.

- The next three RowFilters are used to remove any reviews containing the words *dog*, *cat*, and *pet* to ensure that the reviews are for humans.

- The metanode is a collection of nodes collapsed into a single node; it can be expanded by double-clicking it. The content of the metanode is also listed below the overall workflow. Following are descriptions of these text mining nodes:

  - **PoS tagger**—The first task in any text mining analytics is to tag the individual words in terms of parts of speech. (KNIME uses the Penn Treebank Tagset.)[11] Once the document has

_____
11 http://www.cis.upenn.edu/~treebank

been tagged, a PoS filter can then be used to select only those word types that may be of interest for the analysis.

- **Bag-of-Words (BoW) creator**—The BoW creator takes the entire document and creates a bag of words with their PoS tags and the parts of the document in which they occur.

- **Punctuation erasure**—It is usually advantageous to remove all punctuation marks. When Deep Processing is checked, the terms in the document are also changed beyond just changing the terms.

- **Stop word filter**—KNIME supplies a list of stop words, which are common words that do not significantly affect meaning (for example, *the, a, this, is, was, for, and, some,* and so on).

- **N Chars filter**—Specifies the minimum number of characters of a term. Most terms with very few characters tend to be carry less significance and can be ignored.

- **Number filter**—Removes terms consisting of numbers only.

- **Porter stemmer**—Uses the Porter stemmer algorithm to reduce terms to their stems or roots.[12]

- **TF**—Computes the relative frequency of the term from the total number of terms in the document.

- **Frequency filter**—Can be done either by a threshold value between a Min Max range or by setting the number of largest relative TF frequencies.

- The first tag cloud for just *nouns* shows the reviews consisting mainly of *potato chips (kettle type), coffee, tea, candy* (rendered as *candi* by the use of stemmer within the text mining metanode), and *chocolate* (see Figure 4-14). A selection of non-product nouns are *flavor, product, taste, salt, smell, price, MSG,* and *calorie.* Without knowing their relative

---

12 http://www.eecis.udel.edu/~trnka/CISC889-11S/lectures/dan-porters.pdf

significance, at first glance, the bad reviews are generally about flavors, price, and health-related concerns (salt, MSG, calories).

Figure 4-14   Tag cloud for nouns

- The second tag cloud shows what *adjectives and adverbs* the customers used to convey their feelings about the products (see Figure 4-15). The most frequently used word is *bad*, followed by *stale*, *sour*, *disappoint*, and *strong*, *greasy*, and *mushy*. The seemingly positive adjectives such as *OK*, *fresh*, and *favorite* were modified by negative adverbs or phrases such as *past fresh by date…* and *… is my favorite, but truly this one is not a good buy*. The use of *OK* seems to indicate less than enthusiastic endorsement! They indicate the limitations of simple text mining.

- To obtain more accurate meanings of texts, you have to understand the context, topics, events, and sentiments contained within the texts by the use of NLP and sentiment analysis tools (discussed in the next section).

- **Business applications**—Beyond tag clouds, you can use the results to generate the following:

  - **Alerts** when new words, attributes, and brands (your own and those of competitors) are emerging and forming a strong trend.

- **Attributes** of comments or sites visited of certain group of customers or a section of the social network graph. They characterize the persona of customers or the social network; and improve the prediction of target customer values and behaviors such as wallets, conversion, and clicks.

Figure 4-15 Tag cloud for adjectives and adverbs

- **NLP and sentiment analysis**
  - **NLP approaches**—NLP automates the knowledge discovery process of text documents. There are two main types of NLP: statistical and linguistic. (There are many excellent books and articles on the web that you can browse to gain a deeper understanding.) This section focuses on the applications to business of using the statistical approach with KNIME nodes.
  - **Business applications**—There are many business applications, and new uses are constantly being created. Here are just a few:
    - Machine translation; for example, Google Translator
    - User interfaces such as mobile semantic search

- Speech and sentiment recognition, Siri, call center topics, and sentiment recognitions for more accurate routing of calls to the right department for management and resolution

- **Data set**—The data set downloaded from the University of Illinois Urbana-Champaign website[13] is a data set crawled from the Trip Advisor site.[14] It contains more than 500K of reviews of hotels in the United States with their ratings. To save computation time, I selected the hotel with ID 100560, which has only 64 comments. (In real-life situations, the data sets are likely a lot larger because they span over longer periods and from many more sites.)

- **Workflow**—The workflow, including the data file, can be imported from the zipped file http://bit.ly/Wc3rGl. It reads the Trip Advisor reviews for hotel 100560 and also a sentiment corpus from MPQA University of Pittsburgh (http://mpqa.cs.pitt.edu). The corpus contains 7,651 entries of words and their polarities, and can be downloaded here: http://bit.ly/Wc4cPu. Because it is a CSV file, it can be readily amended or modified to suit the specific context and applications.

The insert in the workflow (as shown in Figure 4-16) is the Configuration screen of the DictionaryTagger node. Because KNIME nodes do not yet support sentiment tags, one way to do it is to assume that the named entities (NE) tag is set as TIME for negative sentiments and as PERSON for positive sentiments, for example.

The text mining metanode is the same as in the previous workflow. By excluding the non-negative comments,

---

13 http://sifaka.cs.uiuc.edu/~wang296/Data/index.html
14 "Latent Aspect Rating Analysis without Aspect Keyword Supervision." Hongning Wang, Yue Lu, and ChengXiang Zhai. 17th ACM SIGKDD Conference on Knowledge Discovery and Data Mining (KDD'2011), P618-626, 2011.

the frequencies of occurrence of the nouns and adjectives/adverbs can be again presented as tag clouds.

- **Business case**—Assume that Joe is the fictitious owner of the particular hotel and that he has started to monitor the customers' comments on Trip Advisor for his hotel on a regular basis. After losing his manager for years to his competitor, a large number of staff members also left. After finding a new manager and replacement staff, Joe was eager to find out what customers are saying about his hotel.

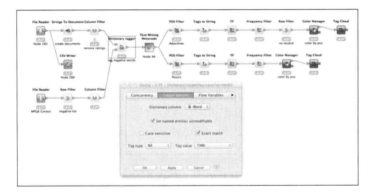

Figure 4-16 Hotel sentiment analysis workflow

- Tag cloud 1 shows the things that the customers in their negative comments were complaining about (see Figure 4-17). They mentioned his hotel by name (masked) and location. They also commented about almost everything in the hotel: *bathroom, lounge, window, pool, service, bed, floor, lobby, breakfast, property, Internet, bar,* and others. Joe was eager to see what went wrong.

- Tag cloud 2 (as shown in Figure 4-18) shows the feeling of the customers and the most frequent terms were *aw** (as in awful) and *bad* (see Figure 4-17). The sentiments expressed by the range of negative words used were quite strong. In addition to the deficiencies in the service of the hotel staff (for example, *rude, unskilled, negative, dirty*), the hotel also needs major updating in lighting (*dark*), furnishing (*out-dated*), and sound insulation (*noisy*).

Figure 4-17   Negative sentiment noun tag cloud

Figure 4-18   Negative sentiment adjective tag cloud

Conclusion

The main lessons you learned from this chapter include the following:

- **Be proficient with the tools**—First is to be hands-on; then practice, learn, and perfect your skills. Never be an analytics leader who has never cooked or stopped cooking with analytics.

- **Different types of analytics**—There are four types of analytics in three stages: Descriptive and diagnostics analytics are for looking at the past and the present; predictive and prescriptive analytics are for looking into the future and being prepared for the future, respectively. Analytics 1.0 is mainly for looking at the past and the present, predictive analytics is analytics 2.0, and what is emerging is analytics 3.0 (prescriptive analytics).

- **Unified knowledge discovery approach**—To avoid being lost in translation and the formation of knowledge silos, BI visualization and reporting tools must be used together with the more advanced analytics tools. Instead of a solid separation between business and knowledge users, there should be a new class of *knowledge business users* who can actively engage and contribute across the functional and organizational silos. Such users must be proficient in using analytics tools.

- **Analytics toolbox**—Besides the BI tools, advanced analytics are grouped into four major categories: tools for predicting likelihood; grouping of people or entities; predicting quantities; and other types of advanced analytics, including text mining, NLP, and sentiment and social analysis.

- **Hands-on analytics modeling experience**—By now, I hope that you have downloaded KNIME and built the example workflows for the various cases listed in the chapter.

- **Likelihood prediction**—You can build simple models for predicting the likelihood of the way a target event or behavior will happen and be applied in business cases.

- **Model training and validations**—You learned the various ways to build predictive models with training and validation samples. The concept of lift charts in evaluating how well a model performs was also explained, with their business implications and tips to avoid misinterpretation.

- **Grouping or segmenting customers**—There are two ways of grouping customers described in this book for learning their group characteristics: decision tree model and clustering by x-means.

- **Simple rules for classifying customers belonging to different classes**—The classification of buyers vs. non-buyers, and churners vs. non-churners can be obtained with decision tree models. Different types of errors and their impacts under different business conditions were also explored together with the costs of determining what level of error would be tolerable.

- **Letting the customers group themselves**—To avoid any preconceived biases, sometimes it can be advisable to use unsupervised learning (that is, x-means clustering for finding the optimum number of groups that produce the smallest amount of inhomogeneity within the clusters).

- **Prediction of quantities by linear regression and ensemble regression tree models**—They are contrasted in terms of their accuracies and business implications. The different crucial factors in affecting the outcomes were also explored.

- **Two voice-of-customers text mining workflows**—How to conduct text mining and the NLP of online reviews of Amazon fine food reviews and Trip Advisor hotel reviews were posted for you to download and run. Potential business applications for listening to the voice of the customers and building of timely alerts and to discover product and content attributes were also highlighted.

5

Analytics Decision-Making Process and the Analytics Deciders

"Handle your tools without mittens."

—Benjamin Franklin

Time to Take Off the Mittens

It seems like common sense, but after over a decade of applying advanced analytics in many businesses, I have witnessed many failures caused by businesses handling analytics with their "mittens" on. These mittens can be overreliance on old tools, past experiences, outdated thinking, and attitudes fostered within rigid and restrictive organizational silos. A Forrester survey[1] (Q1 2012) found that many analytics users are still limited by what they did before while they apply analytics, so they cannot reap its full benefits.

The mittens that caused most analytics projects to fumble were often due to businesses misplacing their focus on the following:

- Short-term volume growth metrics instead of long-term profitability
- Campaign lifts and channel response instead of customer-related metrics

---

1 http://www.forrester.com/The+State+Of+Customer+Analytics+2012/-/
E-RES61433?objectid=RES61433

- Descriptive summary analysis instead of predictive, scientifically tested, and optimized advanced analytics

One way to avoid such mistakes is to adopt a consistent business process, which I call the business analytics process (BAP). *BAP* is defined as a consistent process through which business objectives can be met and insights executed, and then tested with the best in class data and advanced analytics driving strategies and executions. To address what this means, let's begin by examining the different stages in an industry best practice process known as the Cross-Industry Standard Process (CRISP-DM).

CRISP-DM was conceived in 1996 during the early days of analytics (at that time better known as data mining, hence the DM in CRISP-DM). Teams from DaimlerChrysler, SPSS, and NCR met and wrote the first version of CRISP-DM 1.0 in 2000.[2]

Overview of the Business Analytics Process (BAP)

Even though I start with the CRISP-DM process, in my experience the BAP actually involves a few more steps that are critically missing in the conventional CRISP-DM process.

Similar to the CRISP-DM process, BAP also begins with *business* and ends with *deployment* in six steps. However, just having these steps is not sufficient to ensure the success of an analytics endeavor. You also need to build in six feedback loops and formally establish an analytics sandbox.

The following list describes the six steps of BAP:

1. **Business objectives**—In this step, important business questions are raised, and anticipated successes are defined.

---

2 ftp://ftp.software.ibm.com/software/analytics/spss/support/Modeler/Documentation/14/UserManual/CRISP-DM.pdf

Business and key analytics leaders should *jointly* define the objectives.

2. **Data audit**—Once the objectives are defined, an experienced business analytics team should quickly focus on the kind of data needed for the potential models and conduct a thorough data audit to determine data availability within the existing data infrastructure. If there are issues with the data quality or quantity, the team goes back to the first step to determine with the business how to modify the objectives or to further collect or cleanse more data. This is *Feedback Loop 1* shown in Figure 5-1.

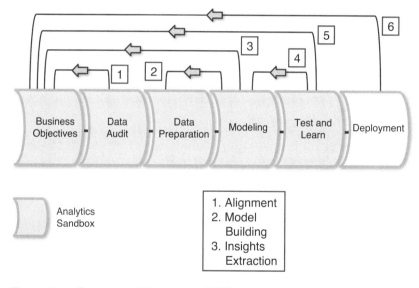

Figure 5-1   Business analytics process (BAP)

3. **Data preparation**—Once data is determined to be of adequate quality and quantity, it needs to be properly prepared into formats that can be fed into models to be built.

4. **Analytics modeling**—The single step of modeling in CRISP-DM is expanded into three substeps in BAP, as follows:

   • **Business alignment**—Before any modeling can be done, the business objectives have to be translated into a series of analytics modeling objectives and briefs. To do this properly, someone who possesses intimate knowledge of both modeling and business strategy must be responsible to lead this step. Unfortunately, because of the shortage of such talents, companies tend to fail in the alignment in two ways. First, prescriptive requests are specified by businesses based on their prior knowledge that essentially limit the knowledge discovery process and over time render the analytics team as simple order takers. Second is a hands-off approach. Because of the lack of analytics knowledge, business owners leave the knowledge discovery wholly in the hands of the modelers, resulting in negative results: 1) Time was wasted; the analytics team took too long over-engineering the solutions. 2) Non-actionable solutions were produced because of the team's lack of intimate operational business knowledge. 3) There were lost opportunities because they discarded insights that might have a major impact on business because the results lacked analytical rigor or elegance.

- **Model building**—After aligning with business, the modeling goals can now be accomplished in actual modeling. As mentioned before, different models can be used to solve the same analytics problem, and they tend to produce results of different levels of lifts and insights. The models to use might depend on the data characteristics, business conditions, and model performance. A series of models is usually built and tested for comparison, and the best-performing model is then used for deployment. During this stage, some data may need to be further modified or augmented. This is *Feedback Loop 2*.

- **Model insight extraction**—The importance of this step cannot be overemphasized. Analytics professionals love their craft (that is, building models), so they typically spend

10 percent of their time in data preparation, 70–80 percent in model building, and another 10–20 percent in extracting insights. Ideally, the proportion of time should be 1/3 spent in data preparation, 1/3 in model building, and 1/3 in extracting insights. During this stage, more questions are often raised, and existing ideas and assumptions are modified as new insights are found. With these new insights, there should be a Feedback Loop 3 to the business to refine, modify, or augment the original business objectives.

5. **Test-and-learn stage**—Once the models are validated, and insights are shared and confirmed by the business, the model insights must be translated into actual business actions with the key drivers and metrics defined. First, the models are tested using a small subset of the data used for actual deployment. If the model insights were not supported by the test results, the models are revisited to determine whether the cause was errors in input data or models, or changes in market conditions or deficient marketing collaterals. This is *Feedback Loop 4*. To carry out the test-and-learn phase, the following steps are needed:

   • **Operational alignment**—During the field tests, all the functional and business units should be involved and trained for the actual deployment. Any potential kinks in the execution should be ironed out during the test-and-learn phase.

   • **Design of experiments (DOE)**—In addition to testing whether the models work in the field as predicted, this is the step in which all the variations in the campaign promotions (for example, creative, copy, offer, fonts, and channels) and any other potential factors might be included into a multicell design. The aim is to find out the incremental effects of each of the levers or factors, and to determine the optimal combination of factors to the Dependent Variables (DV) such as conversion rates, site traffic, or response rates.

- **Scenario planning**—The lever settings and the validated rates can then be fed back to the business strategy team to enable it to build a what-if simulator. The simulator is an invaluable tool for determining what the current investment level and its expected ROI should be, the alignment of all the stakeholders during the deployment stage, and the field-tested reliable inputs to the future planning process. This is **Feedback Loop 5**.

6. **Deployment**—This stage is where the rubber meets the road; all the insights and resources are deployed for real. During the deployment step, each stakeholder should measure and benchmark the results against what was discovered during the test-and-learn stage.

When adjustments are needed, all the relevant connected parts must also be updated. The key issues are ensuring that the infrastructure is stable and scalable to support the deployment, things are occurring as they were tested, and the knowledge and insights generated along the BAP steps are captured and documented for future reuse. The updates and knowledge management constitute **Feedback Loop 6**.

Because of the highly iterative nature of steps 1 to 5, most of today's business infrastructures and organizations are either too rigid or siloed to handle the demands of such workflows. This is where an analytics sandbox should be set up for doing what I call *analytics rapid prototyping*.

Analytics Rapid Prototyping

Because most analytics projects must involve the BAP process at least once, much is unknown when the business objectives are first defined. There is a need to quickly go through the BAP steps once. Often, the requirements for the success of such projects are *agility*, *cost control, constant refinement, robust, and reliable prototypes* that can be scaled to form the core of a full deployment. In essence, it is the Evolutionary Prototyping used in software development. Evolutionary Prototyping is defined as a process to build "a very robust prototype in a structured manner and constantly refine it....The Evolutionary Prototype, when built, forms the heart of the new system, and the improvements and further requirements will be built."[3]

Following the Evolutionary Prototyping framework, the key ideas of analytics rapid prototyping can be defined as follows:[4]

- Uncertain requirements and outcomes in analytics projects.

- Constant iterations and refinements are the norm.

- Agility and speed are crucial to analytics projects' success. Analytics iterations should be in minutes, not days or weeks. Let's assume that instead of solving a complex business analytics problem, you are trying to solve a hard calculus problem in college by dividing it among your friends. Each works on a small part of the problem, and comes back occasionally to discuss progress and intermediate results over a period of weeks or months while you are busy doing other things. Do you think you will get the right answer and also learn from such a process? Unfortunately, this is how most analytics projects are done today!

- In-situ evolution; the solution should evolve through use in its intended environment.

- Core solution; The solution forms the core of the final deployment system.

Such analytics prototyping requires the right teams with the right training working within an analytics sandbox. The best-known uses

---

3 http://en.wikipedia.org/wiki/Software_prototyping
4 Software Productivity Consortium: Evolutionary Rapid Development. SPC document SPC-97057-CMC, version 01.00.04, June 1997. Herndon, VA. Page 6.

of a sandbox, besides being a great place for children to play, is the military's use of a sandbox for planning and exercises.

Analytics Sandbox for Instant Business Insights

As shown in a White House archival photo (see Figure 5-2), President Lyndon Johnson is crouching over a sandbox with his key advisors during the Vietnam War.[5]

Figure 5-2 Sandbox in action

A few observations can be inferred from the photo that helps define what an analytics sandbox should be:

- **Team**—The team of decision makers and their key support staff should be present. I call them the *analytics deciders*. Who they are and what they look like is discussed later in this chapter.

- **Ownership**—The business owner, President Johnson in this case, leads the team. Depending on the project, the analytics sandbox may be headed up by the highest-ranking business stakeholder.

---

5 http://en.wikipedia.org/wiki/File:Lyndon_Johnson_KheSanh_Sandbox.gif

- **Realistic sandbox**—A realistic miniature version of the real battlefield condition is used for simulation, which is where the most reliable predictive models should be deployed.

- **Scenario play**—Brainstorming scenarios, options, and potential risks and outcomes in real time.

- **Information on demand**—All relevant information must be made available on request.

- **Safe environment for sharing**—Risks should be removed from the sharing process. No pulling of rank or other undesirable behaviors should be permitted in the decision-making process within the sandbox. Instead, the data and the outcomes of the scenarios and analytics results form the basis of decision making by the team of analytics deciders.

- **Seamless implementation of sandbox decisions**—The optimized solutions and recommendations should be readily deployable to business within the real-life operational constraints.

With the growing complexities of data and IT data warehouse systems in business, there is an urgent need for the analytics sandbox database. The typical IT infrastructure for supporting well-structured relational database managed by the IT department is not appropriate to support the analytics sandbox in terms of its requirements for agility; and Extracting, Transforming, and Loading (ETL) on demand.

The typical approach usually takes three to six months before the process is turned on, and the data is ready for use. This typical IT-led approach obviously is not suitable for analytics rapid prototyping. By definition, the analytics sandbox database has to be fast, flexible, efficient, free from IT production constraints, failure-tolerant, and cost effective. An analytics sandbox is key to the successful implementation of the BAP. The mandate for most IT departments is to ensure an error-free, zero-downtime, and within-budget IT infrastructure. Flexibility, speed, and innovations are usually not part of IT performance metrics.

The analytics sandbox is depicted as the shaded regions in Figure 5-3.

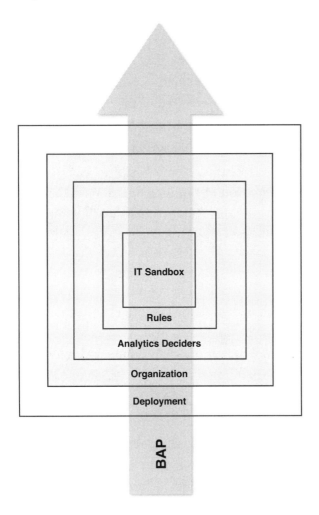

Figure 5-3 Analytics sandbox

Analytics IT Sandbox Database

The characteristics of an IT sandbox database are summarized as follows:

- **For wide and shallow data sets**—Because of different requirements, the analytics data set is usually very wide—that is, with many columns and containing maybe a few 100K rows (an Nth sample of the overall database). On the other hand, the deployment data tends to be much narrower because the unimportant variables have been eliminated and much deeper (for example, for tens of millions of rows).

- **Standalone**—It needs to be a standalone data mart or a separate logical partition in the enterprise data warehouse.

- **MPP and fast technology**—To achieve the real-time and on-demand requirements, the sandbox must use the massively parallel processing (MPP) database together with high-end memory; fastest multicore processors; high-end, high-capacity storage; and input/output (I/O). Examples of MPP databases are HP Vertica, IBM Netezza, Actian MATRIX, Amazon Redshift using ParAccel technology, and Pivotal Software Greenplum.

- **Ready tools**—Favorite business intelligence (BI) analysis and visualization tools and advanced analytics platforms such as KNIME must be made available on the sandbox.

People and the Decision Blinders

In many companies, decisions are made within an organizational silo with data and information supports from functional units such as marketing, sales, and finance. Inevitably, when the timely flow of information is impeded, it can result in wrong decisions.

In a study on *decision blinders*, the authors of "Decisions Without Blinders" in a Harvard Business Review 2006 article[6] observed that "most executives are not aware of the specific ways in which their awareness is limited." They called it *bounded awareness* and defined it as "when cognitive blinders prevent a person from seeing, seeking, using, or sharing highly relevant, easily accessible, and readily perceivable information during the decision-making process." These blinders can have three major effects:

- Failure to *see* information
- Failure to *use* information
- Failure to *share* information

The places along the BAP in which such blinders frequently occur are not within the respective silos. Instead, they tend to occur at the six feedback loops.

Risks of Crossing the Chasms

In many companies, functionally crossing the silo boundaries carries high risks. A businessperson venturing into analytics or vice versa may be viewed as intruding into others' domains of expertise and responsibilities. Hence, this reluctance to incur risks can become a key impediment of seeing, using, and sharing information. Silos formed not just by organizational structures but also by differences in language, culture, and the way successes are measured:

- **Language**—Definitions such as customers, analytics, leads, new customers, loyalty, churn, wallets, values, customer relationship management (CRM), growth, return on investment (ROI), and so on often differ from one part of the company to another. When setting up a BAP, it is prudent to put together

---

6 Bazerman and Chugh. "Design without Blinders," *Harvard Business Review*, January 2006.

a list of common definitions of key terms solicited from and agreed upon by all the stakeholders before the project launch.

• **Culture**—After years of being in business, a business' culture is simply *how things are being done around here*. It is one of the hardest things to overcome for any new innovation that has the potential of upending and disrupting the apple cart. Any executive trying to promote analytics must recognize the challenges and risks involved. It may be prudent to hire a change management expert to help navigate through the legacy culture and organization. It will take time before analytics can be well established enough to usher in new culture and organization.

• **Performance metrics**—It is widely recognized that performance metrics and rewards over time can shape employees' behavior, mindset, and culture. Excellence in analytics must be recognized by its contributions along the entire BAP value chain, not by simply being experts in their own fields. Anyone who tries to cross the silo boundaries and actively fulfill the role of facilitating the various feedback loops must be actively encouraged and formally recognized and rewarded. In my experience, these six feedback loops are where innovations happen. They are, in fact, the intersections in which the Medici Effect[7] flourishes.

The Medici Effect

The six feedback loops are not only crucial for information sharing; they as the intersections of ideas and discipline, are absolutely essential for success. These intersections have long been recognized as fertile ground for new innovations and are where the Medici Effect occurs. The Medici Effect was named after the Medici family who

---

7 http://www.amazon.com/The-Medici-Effect-Elephants-Innovation/dp/1422102823

helped to usher in the Renaissance[8] by bringing together the diverse talents from both East and West.

The author of the *Medici Effect—What Elephants and Epidemics Can Teach Us About Innovation* also pointed out that similar to the intersection in which most traffic accidents occur, this intersection also is "the place where ideas from different industries and culture *collide.*"

Companies rarely reward collisions of any kind because businesses operate better when there are clear lines of responsibilities and domains of expertise. People are also more comfortable staying within their own domains. Unfortunately, if this were the only business structure, there will be little chance for the intersections to exist. By extension, there will be fewer innovations because of the scarcity of the Medici Effect.

Besides the Renaissance, an astoundingly successful example of the Medici Effect is the ultra-successful hedge fund Renaissance Technologies LLC, founded by an MIT and UC Berkeley mathematician James Harris Simons. Renaissance Technologies is renowned for hiring only quants with nonfinancial backgrounds. These quants, which came from diverse backgrounds, were tasked with building mathematical models to detect and capitalize on nonrandom effects in markets across the world. Professor Jim Simons' Medallion Fund outperformed every other hedge fund since its inception in 1989 (averaging yearly returns after fees of 35 percent according to the *Wall Street Journal*),[9] even during the 2008 financial troubles. It also made its founder rich, with a net worth of $12.5 billion and a rank of #94 in *Forbes* list of billionaires.

Instead of viewing the six feedback loops as problem areas to be managed, they should be viewed as opportunities. By actively fostering opportunities from these six intersections, I have witnessed how a

---

8 http://hbswk.hbs.edu/ archive/4376.html
9 http://online.wsj.com/public/health?mod=tff_main or http://www.sarnet.org/lib/simons.htm

process such as the BAP has become the crucible for business innovations. However, not all people can thrive in the intersections. In fact, most people find it uncomfortable or even threatening when they are assigned roles within the intersections.

What kind of people would thrive within the intersections and help bring about the Medici Effect within a company? To become successful, companies must create an environment in which "analytics deciders" can emerge, flourish, and grow.

Analytics Deciders

Such analytics deciders are hard to find and harder to retain. The best retention strategy is to recognize them early and nurture them over time. In my years of finding and having the pleasure of working with such talents, I have found them recognizable by the following traits and interests:

- **Hybrids**—Analytics deciders are likely polymaths with diverse backgrounds and can comfortably move in different circles of expertise. It is usually hard to label analytics deciders and put them within a specific organizational or functional box.

- **Curious about everything**—When James Simons was investigating the moon phases on the stock price, he said he heard about the deliveries at a maternity hospital peaking during full moons. He was curious enough to check and found it not to be true. This anecdote illustrates one of the secrets of his success (besides his passion for business and winning).

- **Good storytellers**—They can communicate well and love to tell stories about technology and business. One of the common challenges I often present to my key team members is to always explain what they do and the analytics insights they find to their neighbors, nieces, and nephews. If they draw a blank from their audience, they need to improve their own understanding of the problem and rework their solutions. One thing I learned early in my career is when answers appear to be too complex and difficult to explain, it usually indicates that the right answer is not found. It is the same in business—when you find the optimal solution, it is often simple to understand and easy to explain and execute.

- **Hands-on**—Analytics deciders draw their creative energy from hands-on analytics modeling experiences. As a result, they never stop doing analytics, even if they become CEOs. James Simons continued to do mathematics research and published frequently long after he became a billionaire!

Any analytics decider who never had hands-on experience or stopped doing analytics risks being an analytics decider in name only. With the growing popularity of analytics, you can find senior executives who boast long lists of analytics skills and experiences on LinkedIn, but have never built a single model and do not know the differences between linear and logistic regressions.

How to Find Analytics Deciders

If the resumes of analytics deciders were to land on the piles of resumes sitting on your desk, what would they look like? Based on the traits mentioned previously and my own experience of reviewing thousands of resumes, the following list shows the traits to look for in analytics deciders:

- **Diverse background**—Analytics deciders tend to be graduates from liberal arts colleges, majoring in social and political sciences or psychology together with heavy doses of statistics, physics, and mathematics.

- **Ability to peel the onion**—They have pursued advanced research degrees (most likely PhDs) in physics, engineering, and other quantitative sciences on topics that require data and analytics skills across many different academic departments.

- **Out-of-the-box experience**—They usually have worked in seemingly unrelated jobs. For example, Simons used to be chairman of the department of mathematics at SUNY Stony Brook.

- **Analytics prowess**—They have proven track records of successfully applying data and analytics.

- **Ability to engage, inform, and influence others**—They enjoy telling stories about the most complex and successful assignments in their careers. They also love to be client facing and are pretty good at it. When interviewing potential candidates, I often ask them to describe their proudest achievements of applying analytics without any visual aids. A true analytics decider can do this in clear everyday English and do it with passion.

- **Quick learners**—They are quick learners of the importance of the softer skills such as people management, sales, leadership, and coaching. Most quants that are not analytics deciders usually sneer at or shun learning soft skills.

It is obvious that given the shortage of analytics talents today, such analytics deciders are hard to find. However, I believe that in the near future, any effective executives and managers will need to be analytics deciders. In fact, any decision makers who are not analytics deciders will likely be an impediment to the success of any business. They hinder the flow of ideas; increase the probability of suboptimal decisions; and create the wrong culture, organizational structure, and reward system in which it is difficult for analytics deciders to flourish.

Becoming an Analytics Decider

If you are someone who has worked within one of the silos, how can you augment your skills and experiences to become an analytics decider? Just as with the Medici Effect, analytics deciders acquire and perfect their skills in the intersections. In the six intersections along the BAP, what they have to do is different, depending on whether they are from the left or right of the loops. The following sections explain.

Intersection 1 (Business ←→ Data)

- **Business**—If you are from the business units, you can become an analytics decider by doing the following:
  - Learn about current data technologies and gain hands-on experience in managing and manipulating data.
  - Be a custodian of customer data; ensure privacy and security.
  - Serve on data governance councils.
  - Organize data workshops for keeping business users abreast of the current data best practices within and outside of the company.
- **Data analysts**—If you come from IT or the data side, you can become an analytics decider by doing the following:
  - Be curious about customers in terms of how they behave and feel about the data collected and what customer insights you can glean from the data.
  - Be curious about business in terms of operations and the entire value chain.
  - Learn to communicate and negotiate with business executives during crucial or critical decision-making moments.
  - Be courageous and skillful when conveying opposing views and recommending changes.

Intersection 2 (Data $\longleftrightarrow$ Modeling)

- **Data analysts**—If you come from the IT or data side, you can become an analytics decider by doing the following:
  - Learn about basic analytics modeling and gain first-hand experience in exploratory data analysis (EDA) and some basic modeling. You can then understand the specifications of the modeling team and may be able to suggest new data sources or combine data into new variables that are more powerful and can be more predictive when used with the current models.
  - Understand the current and emerging data needs of analytics. As more data becomes available, it is important for data analytics deciders to anticipate what data can enrich the current data collections and drive new innovative analytics to create new knowledge assets.
  - Be the custodian of customer data privacy and security.
  - Anticipate data needs during the deployment of analytics modeling insights in terms of stability, speed, scalability, security, and accuracy.
- **Data scientists**—If you are a quant, you can become an analytics decider by doing the following:
  - Learn to pull data. There is a special understanding and appreciation when you pull your own data. It can mean the difference between building a great model and a mediocre one.
  - Keep abreast of trends in new data technology and sources. More new data is being generated and collected than ever before. Most data is either poorly structured or unstructured, and needs special technologies to manage and use it. Knowing this means that you can integrate and enrich your current

data and improve your ability to better predict and serve
customers.

- Maintain line-of-sight to execution. Stop trying to show off
  your smarts. Focus instead on delivering values and enabling
  the ability of the execution teams to win.

Intersection 3 (Business ←→ Modeling)

- **Business**—If you come from the business side, you can become
  an analytics decider by doing the following:
  - Learn about advanced analytics by doing some actual model-
    ing, as laid out in this book. You don't need to be an expert;
    you have to know only enough to function effectively in the
    intersection.
  - Maintain your curiosity about knowledge discovery using
    analytics and participate in the analytics insights extraction
    stage with other analytics deciders.
  - Be an active contributor for representing business needs and
    perspectives in the analytics sandbox.
  - Master the art of innovation with data and analytics insights and
    transform them into unsurpassable business opportunities.
- **Data scientists**—If you are a quant, you can become an ana-
  lytics decider by doing the following:
  - Be passionate about business. If you are not interested and
    passionate about business, the intersection is not the place for
    you.
  - Maintain a line-of-sight to business. While working on analyz-
    ing data or building models, it is important you maintain this
    line-of-sight. Keep asking these questions: where the data is
    produced in the business process, how to explain the model
    results in business terms, why the business should know the

results, what actions the business needs to take, and the benefits that will ensue.

- Get closer to customers and executions. I always encourage my analytics team to take every opportunity to listen to the voice of the customers (for example, by double-jacking in a call center's outbound or inbound calls with the customer service reps). Seeing how things are executed also help you decide what insights are easier to implement and how to make the seemingly "esoteric" analytics insights actionable.

- Learn the softer skills. Trust me; they are much harder than the hardest skills you picked up from all the math and physics you learned in college.

- Shift focus. Stop trying to show off your smarts. Focus instead on delivering business values by enabling the execution teams to win.

- Learn to communicate and negotiate with business executives during crucial or critical decision-making moments. Improve your skills in PPT decks both in preparing and presenting them.

- Be courageous and skillful when conveying opposing views and recommending changes based on solid and tested analytics evidence.

Intersection 4 (Modeling ←→ Testers)

- **Data scientists**—Because you will provide instructions for the field testers to implement, you should know the details about the testing procedures, potential breakages, impacts on business, and ways data is to be collected and processed before becoming available for DOE analysis. You should also try to educate and prepare the business decision makers for receiving and using the results from the multicell tests.

- **Testers** should understand the significance of control groups and ensure that different versions of the test cells are rendered and executed properly. If potential problems are suspected to have occurred, testers must alert the modeling team immediately to ensure the integrity of the tests.

Intersection 5 (Test and Learn ←→ Business)

- **Data scientists**—As an analytics decider, you need to do the following:
  - Be aware of business realities. Even though a complete and rigorous DOE is analytically desirable, it might be too costly or disruptive as a first option. An analytics decider should try to work within the given business reality and construct a good-enough solution that gives dependable and useful results, even if not perfect. If needed, Propensity Score Matching (PSM) analytics can be used to augment the DOE results. Also, you should start with a best case as the baseline and use DOE to further optimize the results.
  - Be firm against the use of nonscientific tests. An analytics decider must be able to persuade and influence the business to reject the conventional methods of benchmarking against look-alikes or results from different times (for example, the same store last year).
  - Resist pressure to provide numbers that the business expects; instead, let the numbers speak.
- **Business**—To be an analytics decider from the business side, you need to do the following:
  - Know the DOE methodology. Multicell control groups are nonintuitive, so make sure that you exercise common sense and go over the DOE designs carefully. Anything that is

counter to common sense must have some explanations and other corroborating evidence before you accept it as true.

- Provide business guidance on the important levers to test. There are many possible levers, such as promotions, offers, channels, and so on that you can use to influence customer attitudes, perceptions, and behaviors. Unlike the customer-facing business team, the analytics team lacks the same proximity to customers and needs clear guidance from business on what levers they should test.

- Always start testing on a limited set of levers and settings. Regardless of test protocols, you always need a pristine baseline control group that is held out from any treatment and must overlap other test groups under various treatment regimens. Providing executive sponsorship and cover for the analytics team to perform valid tests and to learn is one of the most important functions of a business analytics decider, not just for validating campaigns and determining what levers to use. More importantly, DOE is for knowledge retention and management—much critical knowledge can be lost without it.

- Trust the numbers. After they are properly vetted, you should support the findings instead of your experience and gut feelings.

- Be willing to modify thinking, operations, investments, plans, and goals based on the results of the DOE.

Intersection 6 (Deployment ⟷ Business)

- **Data scientists**—As an analytics decider, you need to do the following:

  - Be proactive in recognizing, sharing, and documenting critical insights and knowledge relating to customers' personas, behavior, and the effective lever settings discovered during the test-and-learn step.

- Leverage customer knowledge assets to influence and inspire key stakeholders throughout the company to be aware, educated, and to use the past insights for leaner and more effective ways of fulfilling their respective roles.

- Be the custodian and champion of analytics knowledge assets.

- Participate in enterprise efforts to set up a proper knowledge management system of analytics insights and transform them into a distinct competitive edge.

- **Business**—To be an analytics decider from the business side, you need to do the following:

  - Appreciate the talents and knowledge generated by the BAP and the team.

  - Evangelize throughout the organization to adopt the insights from the BAP. After reviewing the analytics insights of how to identify high-value customers within the first few minutes of a conversation, a CEO of a major client in the financial industry once commented, "I want everyone in the company, including my secretary, to learn this."

  - Think about how the insights gained might affect the enterprise strategy beyond the immediate campaigns (for example, in the enterprise organization structure, reporting lines, executive developments, and go-to market strategies).

  - Incorporate enterprise strategic planning and invite key analytics deciders as members of the planning team.

  - Dedicate adequate funds for setting up and running an enterprise analytics knowledge management system.

Conclusion

When I decided in 2001 to adopt the email tagline "Less is more with analytics" for my IBM AP analytics team, the spelling checker listed *analytics* as a misspelled word. Since then, *analytics* is not only accepted by all spelling checkers, but it has also become so popular that many companies have added the term to their promotional materials.

I have observed many companies in the past decade trying to add analytics into their competencies and failing miserably. After experiencing similar outcomes in different industries and situations, I started to realize that there was a common lesson to be learned underneath all these failures.

As the analytics leader trying to carry out the CRISP-DM process, I was essentially creating intersections across existing silos. Collisions happen at intersections, and anyone perceived as the cause of collisions is quickly removed or reassigned. Fortunately, as the IBM Asia Pacific analytics leader, I was given the budgetary and profit-and-loss (P/L) responsibility. I was fulfilling a role (without realizing it then): a Medici of analytics.

As a result, safe intersections were created across functional and business units in 14 countries in AP. The Medici Effect happened. It resulted in analytics achieving an incremental ROI of 43:1, and making the sales reps who used analytics to generate leads three times richer (and creating quite a few analytics deciders). A few of them recently remarked to me that those days were the happiest days of their career!

Unfortunately, the happy story proved to be the exception instead of the norm. So in this chapter, I augmented the CRISP-DM process into a process I called the business analytics process (BAP) that deliberately builds the intersections as part of the process. I also added a test-and-learn step to ensure that everyone has a common set of metrics for measuring their performance as one single team. These intersections also require a new methodology called analytics rapid prototyping, which is supported by a technology infrastructure: the analytics sandbox.

Last but not least, the type of people who participate, thrive, and lead in the sandbox are identified as the analytics deciders. They can be from any of the functional or business units, but also have the skills, maturity, and ability to cross the silo boundaries. They show specific traits and mindsets. I gave tips on how to become an analytics decider. I also showed companies where and how to find, recruit, and retain such analytics deciders.

6

Business Processes and Analytics

The purpose of this chapter is to explore the business process landscape of a big corporation and to assess to what extent advanced analytics has permeated those processes.

To do that, I review the information technology supporting such processes and address potential shortcomings. In the end, I provide a hypothesis into possible solutions to add more intelligence into business processes by embedding advanced analytics into them.

At the risk of oversimplifying, Figure 6-1 includes many of the key processes in a big company. The specific processes vary from industry to industry, so what follows is a generalized approach.

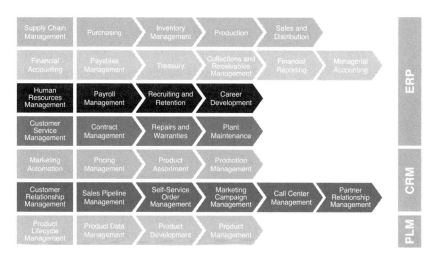

Figure 6-1   Enterprise business processes

I have grouped the preceding processes into seven process families, each supported by one or more types of information technology.

Overview of Process Families

The process families I chose to group the processes into are the following:

- **Supply chain management**—This process family includes purchasing and inbound logistics, inventory and warehouse management, production planning, and physical distribution of goods.

- **Financial accounting**—This process family includes receivables and payables management, treasury, balance sheet, and managerial accounting.

- **Human resource management (HRM)**—This process family includes payroll management, recruitment and retention, and career management.

- **Customer service management**—This process family includes contract management, repairs and warranties, and (for the lack of a better process family to assign it to) plant maintenance.

- **Marketing automation**—This process family includes pricing management, product assortment and promotion management, and campaign management.

- **Customer relationship management (CRM)**—This process family includes salesforce and salesforce management, salesforce automation, e-billing, call center management, and partner relationship management.

- **Product lifecycle management (PLM)**—This process family includes product data management, product development, and product management.

Enterprise Resource Planning (ERP) systems, at different lengths, have the functionality to support the first four process families. CRM systems support CRM processes, hence the name. In recent years, CRMs have grown to support marketing automation processes as well. Finally, there are a number of PLM systems, some of them tightly integrated with the ERP, which facilitate the support of processes related to product management.

I will subsequently describe these systems in more detail, with their capabilities and limitations.

Enterprise Resource Planning

As its name suggests, ERP systems (see Figure 6-2) are software packages for planning and managing company resources, including cash or semi-cash, receivables, inventories, tools and equipment, and human resources (HR). The premise is that an ERP is highly integrated, and various company departments work with the same data from different angles. One of the original goals of the first generation of ERP systems, which became available in the early 1990s, was to reduce the need for systems integration so that different departments would not work on different data sources. As data requirements and available technologies to handle them exploded, the dream of having all IT functionality under one system evaporated. However, ERPs still fulfill reasonably well the aspiration of having company resource data under one system and remain tightly integrated and the one source of truth as it relates to company resources operations.

Among others, the ERP system is the modern version of bookkeeping. Instead of keeping company records in files in cabinets, those records are stored in a central database. Besides efficiently storing consistent records, ERPs are highly efficient at processing operations. You can create a sales order, trigger the creation of a purchase order for the goods you intend to sell, receive the goods into your company, book your vendor's invoice, release payment for that invoice, deliver the goods to your customer, bill your customer, and collect—all from the same system. Other functions that an ERP system can process very efficiently are financial accounting, managerial or departmental accounting, customer repairs and service, plant maintenance activities, production and capacity planning, and many others. It provides a platform for creating the relevant documents that support all those transactions while being very robust at recording all those transactions.

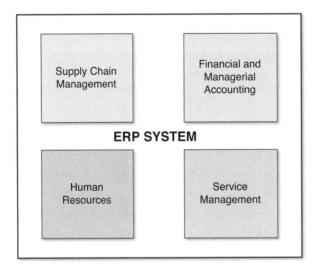

Figure 6-2   Graphical representation of an ERP

An ERP system also supports process coordination and automation, providing significant productivity gains. For example, whenever a non-buying department requests goods or services, the ERP system triggers workflows so that the responsible buyer in the purchasing department instantly receives a notification.

In sales, whenever the company staff creates a customer order, the ERP system checks product availability, not only among the products available in stock but also against outstanding purchase or production orders. By doing so, the system enables the calculation and communication of an appropriate delivery date to the customer.

In production, the ERP system keeps a record of the end-product material and labor structures. Accordingly, when a production order is created, the system evaluates the need for material and labor, calculates product cost, releases automatic requisitions to the purchasing department, and so on.

In many other areas of the organization, an ERP system offers the possibility to configure workflows that significantly automate operations and trigger actions automatically upon configuration.

Many of these functionalities were already available on mainframes, but the prior system generation was less overarching than ERP systems, so companies struggled to maintain data consistency across departments.

Customer Relationship Management

CRM applications (see Figure 6-3) have a much shorter history than ERP systems. The first CRM applications hit the market in the mid-1990s. They started as salesforce automation systems, facilitating the automation of salesforce management tasks. A salesforce automation system enables centralized storage of all information related to sales opportunities, and tracking and classification of sales opportunities along stages in the sales pipeline. It enables sales professionals to record all interactions with customers, to set up follow-up activities, and to organize sales territories and sales management schedules. Additionally, thanks to a CRM application, sales managers are put in a better position to coordinate all efforts among their salesforce and oversee operations within their territories.

Over time, CRM systems grew to include support for marketing, customer service, and support activities as well. Today, CRM applications support online marketing automation and integration, providing coordination of marketing initiatives over different online channels such as email, social media, and blogs. As a result, marketing professionals can better coordinate efforts, track activities, and monitor the results of any given marketing campaign.

Additionally, full modules of CRM systems support customer service and call center management. By recording all interactions with customers, regardless of customer service representatives who interact with them, reps have a complete view of the customer relationship. We all remember the stereotypical interaction with call center reps prior to the existence of CRM systems: a customer called repeatedly, but the call center kept no record of prior interactions. By capturing all interactions with a customer and making the complete history of interactions accessible to all customer-facing employees, CRM systems significantly improve the customer experience as well as call center productivity and efficacy.

CRM systems share some common traits with ERP systems. First, they deliver significant productivity gains by enabling better process coordination and communication. Second, both are systems that mainly support operations and recordkeeping. Third, they operate on top of a single database, providing a unified and consistent view of the data, thus avoiding interfaces between departmental systems.

Figure 6-3 Graphical representation of a CRM system

Product Lifecycle Management

PLM software was developed to build additional functionality on top of computer aided design (CAD) software and provide support to product-development processes.

PLMs include functionality to gain a centralized view of products, and to manage and coordinate unified access to all product-relevant data. They also address the need for interfaces between different CAD software packages and automatic conversion from package-specific to package-neutral file formats. PLMs also facilitate the administration of design changes and change management from proper documentation to approval before release, including workflows to automate such functions. Finally, they enable collaboration among engineers and engineering teams.

Shortcomings of Operational Systems

Operational systems such as ERPs, CRMs, and PLMs do not excel at managerial reporting. Ironically, although these systems store large amounts of data that are very useful for deriving insightful managerial reports, they are not highly efficient at slicing and dicing that data, or at presenting it. For example, ERP systems come with a number of standard reports, but any nonstandard deviation from those reports requires painful changes to the system parameters and sometimes even extra coding.

Some vendors, in particular SAP and its HANA platform, have resorted to in-memory databases to alleviate the pitiful performance of operational systems when it comes to reporting, but other problems exist, including the difficulty to create nonstandard reports, to combine data from outside sources, or to leverage data that does not reside in a relational schema. Finally, the capability of these systems to present the information graphically and flexibly is suboptimal.

At a more technical level, the problem with operational systems and reporting is that operational systems were designed to run on relational databases. Such databases are by design very efficient at creating, modifying, and visualizing full records (rows) regardless of record size (number of columns); but not efficient at running reports, which require the aggregation and visualization of large number of records (many rows), but typically use only a small subset of the columns in a record. To cope with such limitations, the industry came up with data warehouses, which capture snapshots of the operational system data.

However, data stored and organized in an operational system is not in a structure to enable efficient slicing and dicing. That is why IT is required to conduct a number of data transformation steps before business intelligence (BI) systems can efficiently query the data.

The process of extracting data from an ERP system and other sources, transforming the data to be consistent and efficiently organized, and loading the data into new tables from which BI systems will query the data is usually referred to as extract, transform, and load (ETL). This process is lengthy and very costly, but it is necessary to achieve reasonable performance, at least with the technology predominantly available until a few years ago. The final result of the ETL process is a set of tables arranged in a star schema,[1] designed to improve query performance. However, such data structures are not useful as a means to store operational data for use in ERP and CRM systems.

More recently, a new type of database, the columnar database, came to the market. This type of database is significantly more efficient at processing queries involving a huge number of records, and data does not necessarily have to be restructured after it has been

---

[1] There are many excellent books on data warehousing and dimensional design, which is the process of rearranging data for efficient querying. One such book is *The Data Warehouse Toolkit: The Complete Guide to Dimensional Modeling*, by Ralph Kimball.

extracted from its operational source. However, columnar databases are not efficient at the task in which relational databases excel: quickly creating, updating, and retrieving single records (containing only one row but potentially hundreds or even thousands of columns). For that reason, operations and reporting still usually run on different databases.

On top of data warehouses, a number of different vendors provide a BI layer. From previous chapters, you remember that BI is also called descriptive analytics. BI excels at painting a portrait or description of what has happened, both in tabular format as well as graphically in the form of charts and visualizations. Typical questions that such BI systems help respond include these: Which customers bought which products across which regions? Which accounts have bought goods and services over $1 million in the first semester? Which accounts are the most profitable?

To answer such questions, a BI system uses data extracted from ERP and CRM systems, and from many other sources. However, as explained in the preceding paragraphs, for BI systems to efficiently query operational data, such data has to be transformed and rearranged in advance.

If operational systems are ill-equipped to handle descriptive analytics, their handling of predictive or Big Data analytics is almost non-existent. For all its operational efficiency, ERP and CRM systems are not designed to support the process of creating and running predictions. There are very few exceptions (in the forecasting process, for instance), but in those cases, ERP systems perform only very basic time series extrapolation. CRM systems typically include more embedded analytics than ERP systems do—for instance, to route or schedule incoming and outgoing calls based on given criteria. Table 6-1 summarizes the capabilities of the different systems.

Table 6-1 Summary of Operational System Capability

	Support of Operations	Reporting Capabilities	Advanced Analytics
Enterprise Resource Planning	Strong	Poor	Non-existent
Customer Relationship Management	Strong	Moderate	Poor
Product Lifecycle Management	Strong	Moderate	Poor

On the other hand, teams performing advanced analytics work in an information technology environment separate from those of operational systems. Predictions or rules derived through predictive models are usually manually fed back into operational systems. In some cases, connectors exist between advanced analytics environments and operational systems, but the information exchange is not fluent. In other cases, data miners are required to create an XML file that encapsulates the logic of a predictive model so that such logic is executed by the ERP system. However, ERPs can only execute the rules; they cannot learn from experience and subsequently enrich the predictive model.

Embedding Advanced Analytics into Operational Systems

In my opinion, there would be significant advantages if we could find better ways to embed more intelligence into business processes through advanced analytics. To do so, the technology available to process and automate operations should benefit from advanced analytics.

Solving this issue is not easy. The core architecture of some ERP systems is 40 years old, and hundreds of thousands of person-years have been put into its development. Although the situation is not as dramatic with CRM and PLM applications, the foundations were also developed at a time when data mining was still in its infancy. The cost of changing or adapting operational systems to embed advanced analytics is prohibitively high. At the same time, it would be very difficult (although not impossible) for a new player to develop such a more intelligent operational system from scratch because a massive investment would be needed before such a product is ready for operational deployment.

Figure 6-4 is an illustration of a theoretical model for a possible solution. The idea is not completely novel and draws from a very common approach in the IT industry. Whenever a one-fits-all solution is not feasible, we resort to gluing together best-in-kind solutions.

To further elaborate on this idea, all operational systems can be logically grouped into an operational layer. In turn, the term *advanced analytics layer* is used for the software package or packages providing advanced analytics.

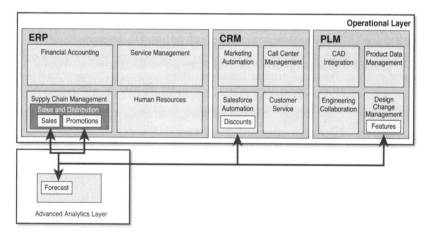

Figure 6-4   Advanced analytics layer

What is needed are efficient connectors or bridges between both layers. An efficient connector is one that is aware of the data schema on both ends, and can automatically and regularly transport data in both directions.

Once the data arrives into the analytics layer, the existing models should be recalibrated and the new resulting predictive rules pushed back into the operational layer, so that processes supported by operational systems can be executed according with the most up-to-date predictive rules. An alternative solution is to use the analytics sandbox to do the Extract, Load, Transform (ELT) on demand. The analytics system stores the massive "raw" data using the NoSQL format and performing the ELT outside of the ERP system. Once the ELT rules and analytics models are built and tested, they can then be embedded within the ERP and run automatically whenever the databases are updated.

Note that recalibration can happen automatically or semiautomatically for minor changes to the predictive model. When more significant changes are necessary, human intervention is needed most of the time.

The number of potential connectors can be as many as the processes in need of predictive intelligence. The best way to convey this idea is with examples, so three are shown here for illustration. All the examples share a common theme, the forecasting process, but each has a different perspective.

Example 1: Forecast

Someone in sales or sales administration is usually in charge of preparing the sales forecast, and sometimes an entire team is devoted to this task. This person or team usually extrapolates from past sales time series. Many quantitative methods exist to smooth out month-to-month quantity variations and to adjust for trend and seasonality. More recently, some practitioners have built complex regression models that compute demand as a function of relevant variables. Examples of those variables include promotions, advertisements, prices, and competitive prices. There are a number of very successful stories built on these types of forecasts.

However (and regardless of the method used), it is not unusual for the internal departments that depend on the forecast to conduct key value-creating activities such as production planning or purchasing to perform manual adjustments on the sales forecast or create their own forecast version altogether.

I believe that very few companies have implemented the idea of a forecast based on a continuous and automatic or semiautomatic data exchange between operations and advanced analytics systems, however.[2]

The way to proceed is to build an analytics-based forecast, in which a function of demand at the end-customer level is created based on the known factors that affect customer demand. In the consumer-packaged industry, for instance, the relevant variables could be demand at point of sale (POS) for the same week in the previous years, prices, competitor prices, advertisement investment, promotional discounts, promotional activities, in-store merchandising, and others.

A different model should be built for different channels (for example, convenience stores, grocery stores, membership-only retail warehouse clubs, pharmacies, etc.) per stock keeping unit (SKU). The company should also model the demand from end customers to intermediaries, and subsequently the resulting demand from intermediaries to the consumer packaged goods company.[3]

At an extreme, a company that sells directly to consumers or businesses without intermediaries could model the demand at the individual customer and SKU levels, as well as the demand for new customers.

---

2 Even though the forecast can be recalibrated semiautomatically using the most up-to-date data from the ERP (that is, the coefficient of the models can be updated or, in analytics terms, the model can be retrained), a significant change to a forecasting model still requires human intervention.

3 An excellent book on the analytics-calculated forecast is *Demand-Driven Forecasting: A Structured Approach to Forecasting*, by Charles W. Chase.

Through a connector, all relevant data to model and create the forecast is extracted from the supply chain or ERP system and then fed into the analytics system. Once the data is in the advanced analytics system, the process of building the forecast model can begin. The end result is a forecast at SKU level, which is transferred down to the operational system through the connector. The operational system, in turn, uses the forecast to adequately plan for demand.

As reality kicks in, the operational systems start getting input on data for variables used in the advanced analytics model (for example, advertisement levels or competitive prices). Through the connector, those variables can be fed into the advanced analytics system; the forecast model should be retrained, the validity of the new forecast checked, and the forecasted sales quantities updated and sent back to the operational system. Once the data is in the operational system, planning can be adjusted if necessary.

Example 2: Improving Salesforce Decisions

To continue with the forecasting example, one of the variables or attributes that has a considerable impact on revenues is discounts granted to customers. Assuming that the analytics team has been able to model such an impact on revenue accurately, and that the forecast was calculated with a pre-agreed-on discount level, changes of discounts are needed in the daily battle for market share.

Some companies grant their sales executives some leeway to grant discounts within a predefined range and put a higher priority on quota achievement than on absolute discounts. In such a situation, it is perfectly possible and admissible that the forecast is not met—not because of a miscalculation but instead an unforeseeable change in premises.

If the event described in the previous paragraph cannot be avoided, perhaps a two-way connector between the analytics layer and the CRM system is possible. The analytics system can push down the forecast model parameters to the CRM system, and a module in the CRM system can provide salespeople with simulations of the impact on revenues because of their discount decisions. Salespeople might change their decisions as a result, but if they keep them, such new data on discounts can be sent up to the analytics layer for recalculation of the forecast. In turn, the connector described in the previous example feeds the new forecast into the ERP or supply chain system, which calculates a new level of demand and adjusts the requirements for raw materials, manpower, and capacity accordingly.

Example 3: Engineers Get Instant Feedback on Their Design Choices

Analytics teams can sometimes build models of end-customer demand as a function of certain product features. Those features are the same as those on which design or engineering teams work. With a proper connector between the analytics layer and the PLM layer, it is possible to adjust the forecasted demand for the next few months based on changes planned by the engineering team. Following a similar logic as in the previous example, engineering teams can instantly view simulations or what-if scenarios of revenue impact upon the implementation and release of certain feature changes. In turn, once those feature changes are final and released for production, they can be fed up to the analytics system for the purpose of recalculating the forecast. The connectors between the analytics layers and the operational systems can then propagate the new forecast figures downstream toward the operational systems so that adjustments are made where needed.

Conclusion

Most companies rely heavily on information technology to support their business processes. Such information systems include the

ERP, CRM, and PLM systems, which I call operational systems. The main task performed by such systems is to automate record keeping and facilitate process automation and coordination. They also provide the data for descriptive analytics or BI. A BI system, in turn, enables executives to better understand what happened in the past at the operational level.

However, operational systems provide virtually no advanced analytics capabilities, nor can we realistically expect such capabilities to be built into operational systems any time soon. Such advanced analytics is usually performed in a separate environment, and although the output of advanced analytics is taken into account to adjust strategies and execution, a tighter integration between the output of predictive analytics and operational systems is needed, with the final goal of embedding more intelligence into business processes.

The irony is that an ERP or CRM system implementation for a big company can cost tens of millions (sometimes even hundreds of millions) of dollars. Although the productivity gains enabled by process automation and quality gains achieved through robust and consistent bookkeeping are significant, there is also a great opportunity to improve ROI significantly with a modest investment in better integration between advanced analytics and operational systems.

In my opinion, one way to capture those value-creating opportunities is to build standard connectors between the operational and the advanced analytics systems. A number of examples of doing this was presented, and you are encouraged to come up with your own list of further areas of integration. Another option is to do all the analytics within the analytics sandbox and then embed the intelligence within the ERP to be deployed in real-time and automatically.

The goal is ultimately to enable an intelligent operational system that can communicate and transfer relevant data to the analytics layer on a permanent basis, and receive the resulting insights from the analytics layer and execute accordingly.

7

Identifying Business Opportunities by Recognizing Patterns

"How could you know which customers will buy…Did you have a 'Crystal Ball'?"

—An IBM sales leader

Patterns of Group Behavior

When you see the effectiveness of analytics predictions for the first time, it does feel like magic. How can you predict which customers would buy certain products? It is like trying to predict how pollen suspended in water would move over time. It is well known that it is virtually impossible to predict the exact movement of a single pollen suspended in water. Albert Einstein was the first to understand the Brownian motion of a single pollen in 1905 that led to a Nobel Prize in 1926 for the experiment that verified Einstein's theory. However, predicting the ensemble or group behavior such as the diffusion pattern for a lump of pollens is actually fairly straightforward.

Likewise, though it is impossible to know with certainty what an individual consumer will do next, it is quite another matter to predict what and how much a group of like-minded customers will buy. This is because groups of consumers tend to exhibit certain *patterns* as a group in terms of what, how much, and why they buy.

This chapter focuses on three types of patterns: *group patterns*: how to group customers, behaviors, or any other entities and be able to detect their underlying patterns; *purchase patterns*: from transactions to detect product purchase patterns (for example, market basket analysis to determine how shoppers would purchase related products if given the chance); and *temporal patterns*: how to leverage past patterns of the target Dependent Variable (DV) or other Independent Variables (IVs) to predict future values of the DV. Before diving into the details, let me share from my experience why being able to discern patterns is important to business.

In this chapter, you see how you can leverage analytics to recognize and monetize patterns in customers' behavior, attitudes, demographics, and over time.

Importance of Pattern Recognition in Business

At IBM, one of the first applications I worked on helped identify the "low-hanging fruit" that can be realized immediately with multi-touch campaigns including email, print mail, and telesales. (Because this is an actual case study, I have disguised the units and country to which it was applied.) Among the small and medium business (SMB) customers, a group was found that used to buy a lot from IBM (in the tens of millions of U.S. dollars per year) for many years, but almost completely stopped buying in the prior 2 years.

Before trying to reactivate these contacts with campaigns, it was decided to use additional analytics to find how responsive they would be when touched. The normalized response rates of the contacts as a function of the # touches per contact per year for the individual business-to-business (B2B) customers is shown in Figure 7-1 (for illustration only). It does not show any discernable patterns that can be leveraged for insights.

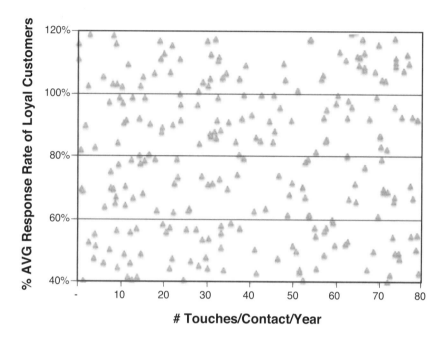

Figure 7-1   Normalized response rates vs. average touches per year

The overall customer segmentation based on past purchase patterns and firmographics was then used to group the contacts. There were five segments altogether found among these SMB customers. When their respective segment averages were normalized, response rates were plotted together with the most loyal customer segment against their average #touches/contacts/year, and a clear pattern emerged, as shown in Figure 7-2.

These SMB customer segments are now grouped into three distinct groups: one representing the loyal customers segment and two other groups of churned segments. The lower two segments belong to a group that was labeled Lost Customers because they were touched less but also responded at lower rates indicating lower interests.

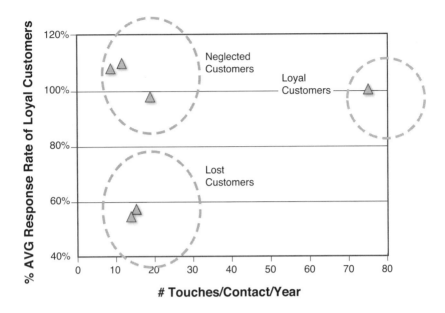

Figure 7-2 Discerning the customers

The most interesting group is the Neglected Customers, as shown in Figure 7-2. Even though these customers were touched at one-seventh of the frequency of the Loyal Customers, they responded at rates similar to the Loyal Customers each time they were touched. In fact, during subsequent outbound calls, these Neglected Customers complained to the IBM telesales reps that they had not received any updated communications from IBM.

The analytics was done near the end of Q3, and two additional outbound calls were placed to these Neglected Customers during Q4. The results in Figure 7-3 showed that more than 50 percent of the lost revenues were recovered in Q4 of the same year. Within 2 years, almost 80 percent of the lost revenues were recovered.

These customers were neglected because most telesales are given more accounts than they can call on a regular basis. As a result, they tend to develop their own rules of prioritizing the leads. One of the characteristics of these Neglected Customers is that they are from smaller accounting or professional services accounts that stopped buying hardware from IBM. In my experience, as more touches of customers are shifted to marketing and customer service call centers, analytics should be used to prevent bounded awareness, as explained in the previous chapters concerning marketing and sales staff.

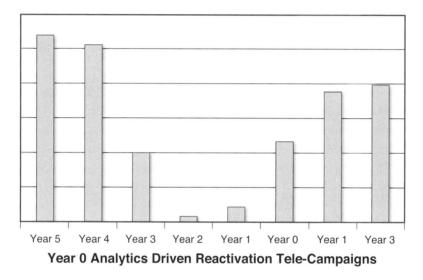

Year 0 Analytics Driven Reactivation Tele-Campaigns

Figure 7-3   Revenues of neglected customers

Group Patterns by Clustering and Decision Trees

As described in Chapter 4, there are a few ways to group and analyze customers based on their characteristics. Again, the simple case of riding-mower ownership is used (in this case, using incomes and lot size): 1) the conventional way of splitting the customers based on internal considerations, along a fixed value of one of the factors and view how similar the respective homeowners in each of the groups are; 2) using a classification tree, which uses multiple splits starting with the most significant factor until a series of more or less homogeneous groups are obtained; and 3) using clustering, which allows the data to group itself based on characteristics.

Because the first conventional method of splitting by products, regions, spends, or a few demographic factors such as gender and incomes have been used as segmentation, it is worthwhile to spend time on the differences between these three methods. Please don't feel intellectually insulted by the simplicity of the case; it is purposely done by design. My goal is to make sure you focus on pertinent business insights, not on large quantities of data and factors. These complexities tend to distract and do not add any more understanding to what I want to cover. You are welcome to try the workflow with your own data set and actual cases from your business.

Three Ways of Grouping

To compare and contrast the way and how well each of the three previous methods performs, you need to define an important metric: the impurities (number of owners wrongly included in a group for non-owners or vice versa). The lower the degree of impurity, the better is the grouping method. However, it needs to be balanced with the number of groups. If there is no limitation to the number of groups, the ones with the lowest degree of impurity (zero impurity) are the groups each has only a single item.

Gini Index of Impurities

The impurities usually measured by the Gini index of impurities are defined as follows:

$1 - (\%$ pure if it is a non-owners group$)^2$
$- (\%$ pure if it is a owners group$)^2$

- **Pure group**—When all homeowners in a group are either owners or non-owners of riding mowers, the Gini index $= 1 - (8/8)^2 - (0/8)^2$ or 0. Hence, a Gini index of 0 indicates the absence of any impurity.

- **Heterogeneous group**—The worst-case scenario is that the group consists of even numbers of both, which results in a Gini index of $1 - (4/8)^2 - (4/8)^2$, or 0.5. Therefore, the Gini index varies from 0 to 0.5, and the key of any split or segmenting into homogeneous groups is to reduce the overall Gini index.

Conventional Grouping—Split Along Constant Annual Income

By splitting at annual income = $59.5K gives two groups in which the Gini index of impurities for the entire data set is lowest:

- For the left group, the Gini index = $(1 - (7/8)^2 - (1/8)^2 = 0.2188$.
- For the right group, the Gini index = $(1 - (5/16)^2 - (11/16)^2 = 0.4299$.
- The Gini index for the entire data set is defined as a weighted sum of the Gini indexes for the two groups: $(8/24)*0.2188 + (16/24)*0.4299$, or 0.3594.

Business Implications

Many customers are incorrectly classified. These impurities reduce the effectiveness of marketing and sales efforts, and affect the accuracy of test-and-learn designs of experiments (DOEs) aimed at learning how potential levers affect the targeted outcomes.

Conventional Grouping—Split Along Constant Lot Size

Let us split the entire dataset along lot size = 18,800 sq. ft. into two groups, one above and another below (see Figure 7-4).

Their Gini indexes are $1 - (9/12)^2 - (3/12)^2$ and $1 - (3/12)^2 - (9/12)^2$, respectively, and are equal to 0.375. Because the two groups are of the same impurities, the overall Gini index is also 0.375, so splitting by income gives a lower Gini index indicating a more homogeneous result than splitting by lot size.

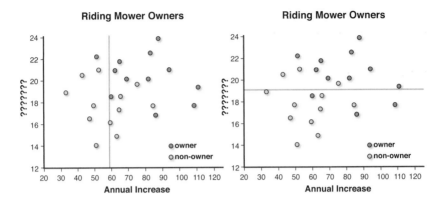

Figure 7-4 Conventional segmentation by income and lot size

Grouping by Decision Tree

By using the simple decision tree model as shown in Figure 7-5, you can do three splits (the first along Income = $78K, the second along Lot_Size = 19,800 sq. ft., and the third along Income = $57.15K) and obtain four groups, as shown in Figure 7-6. The Gini index for all four groups is now at 0.2025 (I encourage you to try to compute this for yourself).

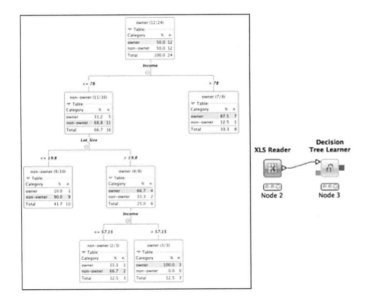

Figure 7-5 Decision tree model result

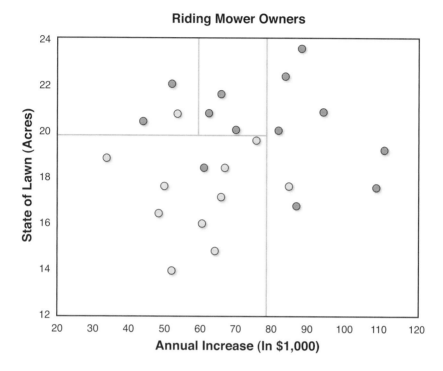

Figure 7-6 Decision tree segmentation

Grouping by XMean Clustering

Use the same workflow from Chapter 4, but setting maxItera-tions = 5, seed ="1234", and range of clusters as between 3 and 5; and cluster the three IVs, including the ownership (you have to transform the ownership into 1 = owner and 0 = non-owner using the rule engine node as shown in Figure 7-7). The KNIME workflow for the entire XMean clustering model is shown in Figure 7-8. Without any computing, the Gini index as shown in Figure 7-9 is clearly zero. The clusters found are purely homogeneous clusters.

From these results, you can see that the conventional grouping has the highest Gini index. Although the grouping by decision tree can produce purer groups with more splits, XMean clustering produces the purest groups with zero impurities.

Expression

```
?  1  // enter ordered set of rules, e.g.:
?  2  // $double column name$ > 5.0 => "large"
?  3  // $string column name$ LIKE "*blue*" => "small and blue"
?  4  // TRUE => "default outcome"
↓  5  $Ownership$="owner"=>1
↓  6  $Ownership$="non-owner"=>0
```

Append Column: Owner?

Figure 7-7   KNIME rule engine node configuration

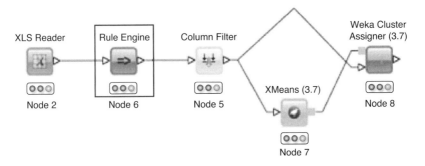

Figure 7-8   KNIME clustering workflow

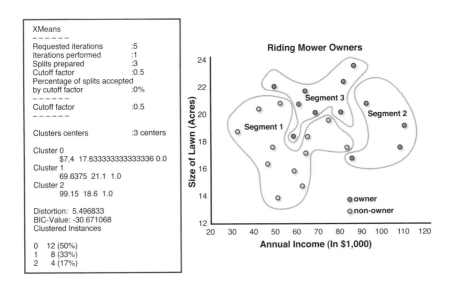

Figure 7-9   XMean clustering result

Recognize Purchase Patterns: Association Analysis

Apart from grouping customers and things, insights can also be gleaned from the patterns of what people buy together. Employing the market basket analysis or affinity analysis can answer this question: "Given that the customers bought Product A, what are other products would the same customers most likely to buy at the same time?" From the current purchases the customers put inside their shopping baskets, you can find the most profitable combinations to ensure that the slower-moving products get to "ride on the coattails" of the precedent products.

Association Rules

In this section, we examine the basics of association analysis and its business implications. The aim is to make sure you can grasp the definitions and basic premise of the analysis and hopefully be able to apply them to your own business.

Business Applications

Business applications for association analysis can range from product development to marketing promotions, product offer bundling, next best offers, and retail product assortment and shelf display planning.

Analytics

The transactions or other behaviors of all customers are first converted into bit vectors (that is, a series of 1s and 0s, indicating whether the item is present or absent in the basket, respectively). From the bit vectors, the more frequent combinations are then used for analysis. Such frequent combinations are called *frequent item sets*. Let's cover some basics before you dive into a real case.

The basic notation of association rule for the frequent item sets is A→B. The rule is this: given A, then B (A is known as the *antecedent*, and B is the *consequent*). It is important to remember that A→B is *not* equal to B→A.

A simple illustration is this: when customers buy a smartphone, they also tend to buy a screen protector: Smartphone→Screen Protector. However, the reverse is not the same. The rule Screen Protector→Smartphone (customers buying a screen protector also buy a smartphone) is less likely to occur. They must already own smartphones; otherwise, they would not be buying screen protectors. So the chance they will be buying another smartphone is low unless they are buying for someone else.

There are a few important definitions before you start on a business case:

- **Support of A & B**—*Support* is defined as the probability of A and B occurring in the entire transaction records. For example, if bread and milk appear together in 1,000 point-of-sale (POS) receipts out of a total of 100,000 receipts, you can say P(Bread&Milk) = 1,000/100,000, or 0.01. It is usual to set a minimum threshold for support; low support means that the rule is not very useful because few customers actually buy both items together.

- **Confidence**—*Confidence* is defined as the conditional probability of the consequent given the antecedent, or P(Consequent | Antecedent). Confidence here is unrelated to confidence level or a confidence interval in statistics.

  Recalling the Bayes theorem for conditional probability that P(Y|X) = P(X&Y)/P(X), you can define *confidence* as the probability of both occurring together over the probability of finding A in a random selection of receipts. This can be expressed as Confidence P(B | A) = P(A&B)/P(A).

Confidence is important. It is the probability of B occurring, given that A occurs. The higher the confidence, the more important is the rule. However, high confidence alone does not mean strong association because when the probability of A and B is high, the confidence will be always higher than P(A&B) as P(A) is < 1 and therefore P(B|A) is by definition also high.

Assume an extreme case: A and B are completely independent from each other (no association). The P(A&B) would simply be P(A). P(B) and the confidence when A and B are unrelated is then P(B). P(B) is referred to as the *benchmark confidence*, or the P(B given A) when A and B are not related.

• **Lift**—The *lift* is defined as how much the confidence P(B|A) would be greater than the benchmark confidence. So the lift is defined as P(B|A) /P(A) or P(A&B)/P(A) /P(B). You can see that if A and B are unrelated, the lift is equal to 1, or there is no lift from using the rule when compared to using no rules.

Business Case

Suppose that a drugstore chain wants to increase sales of some of the slower–moving beauty products, and wants to target its cosmetics buyers. Because cosmetics buyers are among the more affluent customers, the store hopes these high-value buyers will buy these products if the products are appropriately displayed and promoted. The options they are considering include POS specials, cosmetics section end-cap displays, and bundled promotions. (This example uses the data set Cosmetics.xls provided in the XLMiner data mining book.[1])

The data set Cosmetics.xls, which can be downloaded from http://bit.ly/UmPpjV, contains 1,000 cosmetic transactions for customers

---

1 http://www.amazon.com/Data-Mining-Business-Intelligence-Applications/dp/0470526823/ref=sr_1_1?s=books&ie=UTF8&qid=1406068343&sr=1-1&keywords=xlminer

who have bought some of these products: bags, blush, nail polish, brushes, concealer, eyebrow pencils, bronzer, lip liner, mascara, eye shadow, foundation, lip gloss, lipstick, and eyeliner.

Note that there are no amounts, quantities, or customer IDs in this data, so you can't know who bought what and how much. The aim of association analysis is to not identify the purchase behavior of specific customers, but what they tend to buy together in a single shopping trip.

KNIME Workflow

The workflow for discovering the association rules are quite simple: Open a new **KNIME** workflow, which is initially blank. Click and drag the nodes from the nodes repository to the blank workflow editor. The steps are as follows (if you want to simply download the workflow, you can download it from http://bit.ly/1udmyyO and import the workflow into the editor):

1. Read the data file Cosmetics.xls.

2. Before running the Association Rule Learner, the file needs to be converted into a BitVector. As shown in Figure 7-10, the values are used and converted (except for the Trans. #s) into BitVectors. The BitVectors are defined as a string of 0s and 1s, denoting the product not bought and bought, respectively. As shown in Figure 7-11, the first row BitVector is 10000111011110, which represents the transactions Eyeliner (1), Lipstick (0), Lip Gloss (0), Foundation (0), Eye Shadow(0), Mascara(1), Toner(1), and so on.

3. The BitVector is then input into the learner node with the configuration setting shown in Figure 7-12. The minimum support is set at 0.1 to find more rules and to check the output association rules with a minimum confidence of 0.3.

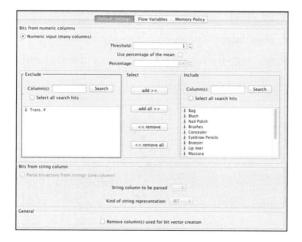

Figure 7-10 Configuration for BitVector generator

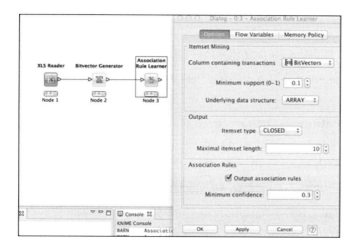

Figure 7-11 BitVector result

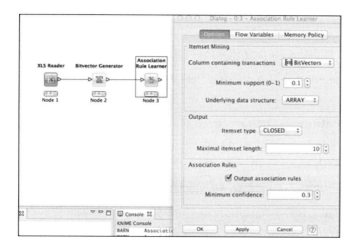

Figure 7-12 Association rule workflow

4. Click the column label lift and choose Sort Descending. You obtain the list of association rules, as shown in Figure 7-13. Let's look at these rules and see how you can make use of them:

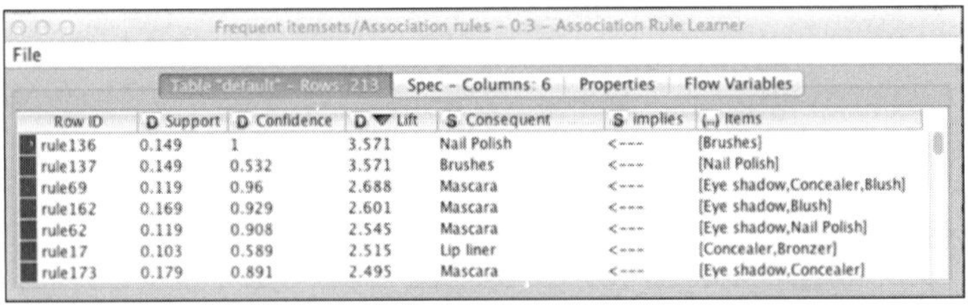

Row ID	Ⓓ Support	Ⓓ Confidence	Ⓓ ▼ Lift	Ⓢ Consequent	Ⓢ implies	(...) Items
rule136	0.149	1	3.571	Nail Polish	<---	[Brushes]
rule137	0.149	0.532	3.571	Brushes	<---	[Nail Polish]
rule69	0.119	0.96	2.688	Mascara	<---	[Eye shadow,Concealer,Blush]
rule162	0.169	0.929	2.601	Mascara	<---	[Eye shadow,Blush]
rule62	0.119	0.908	2.545	Mascara	<---	[Eye shadow,Nail Polish]
rule17	0.103	0.589	2.515	Lip liner	<---	[Concealer,Bronzer]
rule173	0.179	0.891	2.495	Mascara	<---	[Eye shadow,Concealer]

Figure 7-13   Association rule model results

- **Rule 136 (shown in Figure 7-13)**—This rule shows the following:

  - The probability of transactions containing both brushes and nail polish is 0.149.

  - The confidence is 1, which means the P(Nail Polish given Brushes) = 1, or the probability of any transaction including nail polish given it already has brushes is 1 or 100 percent. This is the highest confidence possible.

  - The lift of the rule [Brushes]→[Nail Polish] is P([Brushes & Nail Polish])/P(Brushes)/P(Nail Polish) = 0.149/0.149/0.28 = 3.571. This means that the use of the rule would result in a lift of 3.571 over if you assume that brushes and nail polish are unrelated.

  - **Applications**—The drugstore could offer a coupon for nail polish when customers purchase a brush or simply place the nail polish close to the brushes.

- **Rule 137 (also shown in Figure 7-13)**—Although Rule 137 has the same lift and support as Rule 136, it has a smaller confidence value of 0.532. This means that there is a probability of 0.468 that when customers buy nail polish, they are not buying brushes. The rule therefore is not as robust as Rule 136.

Business Implications

Following are the business implications:

- Choose strong rules that have high enough support (typically more than 10 percent) to prevent promoting items that are not strongly related to each other.

- Choose rules with the highest confidence because the conditional probability of given the antecedent that the consequent is also present must be high (as close to 1 as possible).

- Choose high lift (much greater than 1) to ensure that the most effective rules are chosen, not because of the large presence of the consequent.

Patterns over Time: Time Series Predictions

Businesses need to recognize the repeated patterns of data over time. Retail businesses have long used the same store sales (from the same time period in the previous year) as a way to benchmark their performance. The advantage is that biases in terms of seasonality and stores are usually removed. However, the biases of weather, economic conditions, competitor actions, or any conditions related to the consumers or products unique to just one year may still remain. With rapid changes in the marketplace and consumer vagary in their sentiments and loyalty, such benchmarking is less and less acceptable.

A better way is to build a predictive time series model with all the contributing factors. The historical time series data is used to predict the target quantity (for example, store revenues, number of visitors, profit, costs, and media mix allocations). Again, instead of covering the theory of time series predictions, I will instead focus on the introduction to the business applications by way of actual hands-on examples.

For further readings, the readers are encouraged to refer to the many excellent texts covering the fundamentals of Time Series theory such as auto-correlation, non-stationarity, ARIMA, and VARIMA.

Time Series Models

Examples of time series can be encountered in every business that generates data over a period of time, including stock prices, S&P 500, Forex rates, product prices, inventory levels, revenues and profits, attrition rates, daily temperature and weather data, and so on. In business, time series are often used for forecasting demands and revenues.

For analytics, to illustrate how the time series prediction works, the case of foreign exchange rates of Australian, British, Canadian, Dutch, French, German, Japanese, and Swiss currencies against U.S. dollars from 12/31/1979 to 12/31/1998 is used, as given in the KNIME's 001001 time series example. To minimize the computation time, only a small portion of the overall data is run.

The workflows can be downloaded from http://bit.ly/UwvTRR to first predict the exchange rates of British pounds to U.S. dollars using its past data only. This is a case of auto-regression; by training a model on its past data, a pattern is generated that enables it to predict future values. The second case is to add the past exchange rates of two minor currencies, Swiss and Dutch, to the Independent Variables (IV's). Again, the KNIME workflow with the data can be downloaded from http://bit.ly/WPpHGc.

Predicting Using Its Own Past Data (Autoregression)

The workflow shown in Figure 7-14 consists of the following:

- **Preprocessing**—This metanode deals with missing values and filtering out non-British currencies and normalizing it to within a range (0,1).

- **Lag British rate**—This metanode uses the LagColumn node to produce 100 columns of previous nth day of exchange rates (n is from 1 to 100). So for each row at a certain date, it first contains the exchange rate for that day. The next column in the same row has the previous day's exchange rate; nth columns to the right will be for the previous nth day's exchange rate.

- **Ensemble regression tree model**—A loop based on a table defined by the upstream RowFilter node is used to perform for 100 days of predictions. The tree ensemble regression learner node uses 99 days of the 100-day data to train the model. The learner node is configured using a total of 10 models and a random seed of 123467 at a 70 percent sampling rate for the 10 models.

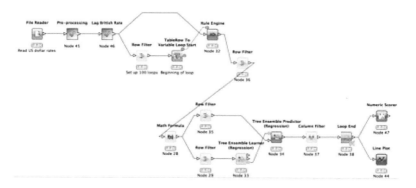

Figure 7-14   Time series Forex prediction workflow

- **Results**—The actual exchange rates and their predictions are shown in Figure 7-15. The overall fit is not too bad, except the predictions tend to lag behind the actual. The numeric scorer gave an R^2 of 0.93, as shown in Figure 7-16, which is very good. Despite this, the score and other typical error measures may not be good measures of performance.

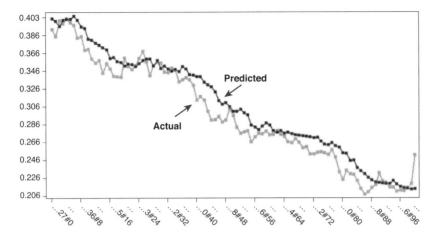

Figure 7-15 Forex predictions from past patterns

Table "Scores" - Rows: 5	Spec - Column: 1	Prope
Row ID	D Tree Ensemble Response	
R^2	0.93	
mean absolute error	0.012	
mean squared error	0	
root mean squared deviation	0.015	
mean signed difference	0.01	

Figure 7-16 Prediction accuracy of autoregression model

The reason why this solution might not be an acceptable solution is that by the next day, the previous day's data would have already been known. So it is *not* a prediction! This is an important lesson: When you look at predictions vs. actual rates involving time series predictions, you should not pay attention only to the shape and trend but also on the time lags in a particular slice of time!

- **Business implications**—Let's try this mind experiment of using the most primitive model possible (that is, the simplest model of equating the current rate with the previous day's rate). This prediction then tracks the distribution perfectly with just a lag of one day! In fact, if the time series do not vary too dramatically from one day to the next, the R^2 and all the other error measures would all be close to perfection. But such a prediction model is totally useless for deciding whether to buy or sell a currency.

A better measure is the timeliness of prediction: how many days the model can predict an impending major event (sharp rise or fall) within a particular threshold of accuracy and confidence! For business applications such as running marketing mix models on the impacts of media spending mix on revenues, the timeliness and sizes of the increases or declines significantly affect the strategic decisions. In the case of Forex trades, the apparent lag of five days of the prediction behind the actual is clearly not acceptable for the purpose of forecasting the impending changes!

- **Exercises**—As an exercise, you might want to modify the workflow to exclude the prior week's data to see whether the time lags can be reduced.

Prediction Using More than One Historical Time Series Data

Instead of using autoregression, let's use two minor exchange rates of Swiss and Dutch currencies into the IVs *without using the past rates* of the British pound.

- **Results**—The results of predicted normalized rates against the actual rates are also shown in Figure 7-17. As expected, the value of R^2 as shown in Figure 7-18 decreased from 0.93 (for the auto-regression case) to 0.728, and the other error measures have also increased because of the inability of the model to predict the initial part of the time series. The model could match the performance of the auto-regression later in the time series. Because no British data was used for the prediction, the time lag problem seemed to have been partially alleviated. In fact, the predictor could predict the sudden uptick by 3 days before it happened! If the model can be validated for consistency, it will be a significant advantage over the competitors if no one knows about it. However, because the workflow is published in this book, the "efficient market" will return to equilibrium by quickly removing this localized advantage!

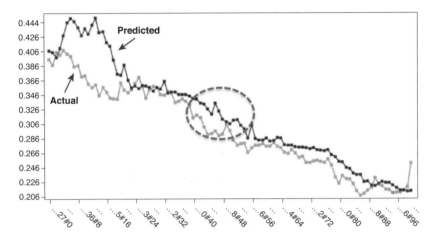

Figure 7-17 Prediction results with other currencies

Table "Scores" – Rows: 5	Spec – Column: 1	Properties
Row ID	**D** Tree Ensemble Response	
R^2	0.728	
mean absolute error	0.021	
mean squared error	0.001	
root mean squared deviation	0.029	
mean signed difference	0.02	

Figure 7-18 Accuracy of prediction with other currencies

- **Business implications**—From the preceding observations, it is noteworthy to pay special attention when using auto-regressions for predictions. It is often best if *expert business knowledge* can be applied to leverage the causal effects of IVs on the target variable or the DV. For example, to predict shopper traffic to a retail supermarket chain, you can use temperature variations, other weather data such as precipitation (rain or snow), promotion and media spend calendars, economic indicators, competitor actions, and in-store experiential marketing. Once the model has been properly calibrated, it can be used as a simulator to aid scenario planning and determine investment levels and contingency plans if adverse and worst-case conditions were to happen.

- **Exercises**—Try to add other currencies one by one and see whether the accuracy and the ability to predict future values work with a certain lead times.

Conclusion

In his *Art of War*, the well-known Chinese General Sun Tse recognized the winning edge afforded by the ability to see patterns that others cannot. In today's increasing complexities and explosive data deluge, the risks of being obfuscated by data are getting higher. It is more and more likely to see a small part instead of seeing the whole, which is like the proverbial blind men touching the elephant. This chapter showed how the analytics methodology can be used to detect and see three types of patterns and how seeing such patterns can affect business outcomes.

The first way of detecting patterns with specific outcomes is done with decision tree models. These outcomes can be buying or not buying, new or existing, best customers vs. others. It is often important for businesses to know what the differences between buyers and non-buyers are, new customers vs. existing customers, look-alikes to their best loyal customers, and other information.

Knowing the differences allows differential treatments of various groups of customers. The decision tree model not only can show you the different patterns and groups related to certain outcomes, but it can also provide intuitive rules that are easy to explain to business owners and non-quantitative execution teams.

When there are no specific outcomes (especially at the beginning of the knowledge discovery process), you might simply want to know the patterns of customers or entities in terms of their characteristics. Clustering models can be used to detect such patterns by allowing similar customers or entities to group themselves for easier detection and profiling. Insights that may be hidden among individual entities may become apparent and actionable in groups or segments.

Once grouped into segments, the customers can be profiled and provided with appropriate and meaningful labels. When there are groups of like-minded people or customers, segment persona can often be inferred by ensemble characteristics. Segment persona is invaluable for experienced media planners to develop compelling and engaging promotions and offerings to entice, activate, and win customers.

Most businesses generate historical data that are not only indicative of trends but can also be used to predict future events. Time series prediction models can be built using time lags and various predictive models such as regression trees, artificial neural nets (ANNs), or support vector machines (SVMs).

It is also shown that conventional metrics for model performance in terms of R^2 and the other error measures might not always be good indicators of performance. By understanding the potential causal relationships between various time series and the target variable, the causal factors can be used to construct effective models to predict the future values of the target variable.

8

Knowing the Unknowable

"Where there is an unknowable, there is a promise."

—Thornton Wilder

Unknowable Events

History is replete with examples of how seemingly unknowable events in history were ignored or feared by most, but leveraged instead by a few as promising opportunities. David Ricardo, who lent money to the British government for the Battle of Waterloo before knowing the outcome of the war is one such example. Another is Warren Buffet, who sold earthquake insurance to the state of California earthquake authority as alluded to in a *Capitalism and Society* article by Richard Zeckhauser titled "Investing in the Unknown and Unknowable."[1] Obviously, after events occur and outcomes are known, these events are no longer unknown and unknowable (UU) or unknown, unknowable, and unique (UUU).

There are still plenty of events in business that are still viewed as UU or even UUU. However, whether they are truly unknowable is unimportant. What is important is that by applying business analytics on the underlying data, you can acquire some of the skills that

---

1 "Investing in the Unknown and Unknowable," Richard Zeckhauser, *Capitalism and Society*, Volume 1, Issue 2, 2006. Article 5, Harvard University.

Zeckhauser surmised are requisite skills to deal with the UU events (for example, decision theory and complementary skills).

In this age of rapid technology changes in data and analytics, the complementary knowledge that today's Ricardo or Buffet would wield would be defined by their analytics prowess. The complementary knowledge would come from their abilities to consume data by the creative use of advanced analytics better than anyone else. There is also something Ricardo and Buffet could not do, but can be done quite easily today: to use extensive data to train and validate predictive analytics models, which can then be tested in the field for causal levers and optimized before deployment. Describing this is the purpose of this chapter.

Unknowable in Business

In the years spent applying analytics in business, I have encountered quite a few people (from business units or sometimes even the analytics team) who tried hard to convince me that trying to understand some UU events were futile. Many felt so strongly when I persisted that they became agitated as if some inviolable dogma had been violated.

Common examples of these unknowables may be "you can't build any useful models in China because the data is too poor"; "you can't predict whether a customer will do something or not"; "you can't cluster stores on nominal or ordinal variables"; or "you can't know the true wallets of the customers if they buy little from you." And there are many others. The readers can bring their own examples from their own experiences. One tip: these naysayers usually begin their sentences with "it can't be done." My answer in these cases is simple: "let's try it and see."

With today's analytics tools and sandbox, you can spend a few hours to see whether it is truly UU. Worst case is that you waste a few hours. But I have found that such wasted pursuits often are the most enlightening because they bring you into areas where few have ventured before. So rather than trying to convince others, I was often the one to roll up my sleeves and try it. I had the excitement of being the first in "knowing" the unknowable. All the examples in this chapter are first-hand discoveries from such quests.

Many non-UU events are often considered unknowable when one or more of the following conditions are encountered:

- **Poor or inadequate data**—The quality of most business data is never perfect; it can contain significant missing data, a limited number of IVs, duplicated accounts, and/or entry errors. As a result, many businesses prefer to clean up or get more and better data before attempting analytics. However, even if substantial investments were spent, data quality issues may remain. So how much do you do before you try analytics? My advice is this: Don't wait.

- **Do it**—Take what you have and follow the business analytics process (BAP) methodology to build and validate predictive models for your specific business objectives. Some applications may require pristine data, but most business objectives simply require that 20–30 percent of the data be representative and of decent quality. The model lifts for a series of training and validation runs will inform whether the data set is of adequate quality and quantity in supporting effective analytics models. Knowing this, you can start quantifying the scope, costs, and business benefits of such data-augmentation efforts.

- **Business case**—When I became the AP analytics leader for IBM, one of my first priorities was to apply analytics to business-to-business (B2B) customers in greater China. Before I started, my China team and some of the business owners advised me to hold off from doing analytics and wait till the data was cleansed. The background is that because of the rapid growth of business in China in 2000, the data was not of the best quality. Despite this issue, there was an annual fight to get the business units to pony up their shares of the costs of data cleansing. This is understandable because the businesses do not see any direct benefits from the effort, only an unwelcome burden.

- **Outcome**—I persisted, and the result was a lift over not using analytics of a whopping 700 percent. Using the total predicted revenue loss because of the uncleansed data in the hundreds of millions of dollars, the costs of cleansing were rounding errors. The business unit sales leaders realized the benefits of winning more business, so they helped to ensure that the budget was quickly approved and the data cleansed for further analytics within months.

- **Limited views**—Rarely in business can you get a full 360-degree view of a business situation and data. There are always gaps in the data, but a creative use of analytics and available data can often help you bridge the gaps.

- **Predict individual behavior in real-time**—From past data on customer behaviors under many varying conditions and levers, you may use time-dependent holdouts to predict future behaviors based on past behaviors and lever settings. By profiling and targeting customers in real-time, they can then be accorded the right treatments or levers that match their values and preferences. As long as there is sufficiently rich and varied data, the causal effects of the levers can then be isolated and optimized for costs and impacts.

- **Hard to prove causality**—As the idiom goes, the proof of the pudding is in the eating. Whether a lever has an effective causal relationship with the target behavior can be proven only by suitably designed control experiments.

Poor or Inadequate Data

When data contains significant missing data, there are a few ways to alleviate the impacts or assess the model effectiveness under such imperfect conditions. You can use data imputation (replacing the missing data by average, median of the nearest values, or a prespecified value) or eliminate the columns of IVs with questionable quality. If the model training and validation performances remain acceptable, the impacts of the missing data are clearly manageable. Additional data cleansing and augmentation can then be justified by the degree of enhancement of the model lifts and the business impacts in terms of additional revenues and profits resulted from such incremental lifts.

Another alternative is to eliminate the rows (of customers) with questionable reliability. Such rows may contain duplicate accounts or outdated or erroneous data. If the reduced set can still support models for predicting individual customer values such as spend amounts or wallets, the validated model can then be applied to customers with uncleansed and missing data. Cleansing and augmenting the questionable data will improve the model performance and ensure that these customers can be targeted for marketing and sales campaigns. This process results in incremental revenues and profits that can be shown to easily offset the costs of data cleansing and augmentation.

Data with Limited Views

There are many business applications in which business owners have only a partial view of their customers' behaviors. Most customers do not shop at just one store or browse at only one website. You can use detailed transactions, iris or motion sensors, or wireless beacons or cookies to track the customers' behaviors in your physical or online stores. However, once the customers leave your premises, you lose sight of what they do. Through third-party cookies and ad networks, you may get a glimpse that the visitors coming to your site visited another sites. You can then serve up more relevant ads specifically to target them. However, most of this information is still quite rudimentary and sketchy (privacy notwithstanding), and you cannot do the same with your physical stores. The analytics described in this section can be applied to extend the views of the business on any variables for which business does not have a complete view.

The model to be discussed in this section is what I personally discovered during one of those UU pursuits and subsequently enhanced over the years. I hope you will download the workflow (even though some proprietary information has been redacted), run it, and see how it works. I have successfully applied the technique to predict customer wallets and lifetime values (LTVs) in many projects that showed significant incremental lifts over other techniques in predicting wallets. I called it the buyer differential characteristics (BDC) technique when I introduced it to my IBM team in 2001 because it allows the business to use the IVs to differentiate between the buyers of different values. The target Dependent Variable (DV) may be wallets, LTVs, or individual product wallets. I discuss a case of predicting the grocery wallets of individual households.

Business Case

Mike, the owner of ABC, a mid-sized supermarket chain, suspects that many of the customers who buy little from ABC would purchase more if treated appropriately. However, in order to know what should be done to earn their business, he would need to know who they are and how much they are spending at other supermarkets. To estimate their spending, ABC supermarkets want to predict the wallets for all their customers.

From the shoppers' predicted wallets and their current spending amounts, ABC supermarkets can then estimate their share of their customers' wallets. Coupled with their propensity to buy products from new categories, ABC supermarkets hope to come up with specific offers and programs to entice these high-wallet/low-penetration customers to spend more at the supermarkets. Let's go over the case in steps using the workflow and data as outlined in the following sections.

Model Data

The data is a small sample of 446 shoppers that was model-generated to represent typical shopping behavior. The statistics of the DV and IVs are shown in Figure 8-1. The DV is the wallet, and the IVs include what the shoppers bought in bakery and dairy, meats, and eggs (assuming that they are found to be the most strongly correlated to the wallets). The actual wallets are not used in the predictive model; it uses only the total spend and the purchase amounts of the reference purchases.

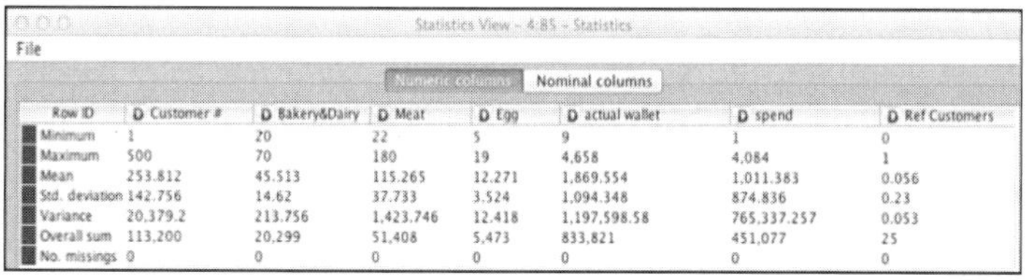
Statistics View – 4:85 – Statistics

File

Numeric columns | Nominal columns

Row ID	D Customer #	D Bakery&Dairy	D Meat	D Egg	D actual wallet	D spend	D Ref Customers
Minimum	1	20	22	5	9	1	0
Maximum	500	70	180	19	4,658	4,084	1
Mean	253.812	45.513	115.265	12.271	1,869.554	1,011.383	0.056
Std. deviation	142.756	14.62	37.733	3.524	1,094.348	874.836	0.23
Variance	20,379.2	213.756	1,423.746	12.418	1,197,598.58	765,337.257	0.053
Overall sum	113,200	20,299	51,408	5,473	833,821	451,077	25
No. missings	0	0	0	0	0	0	0

Figure 8-1 Basic statistics of grocery purchase data

KNIME Workflow

The workflow can be downloaded from http://bit.ly/WalletPredict and is also shown in Figure 8-2. Besides the miscellaneous nodes, there are three major parts of the workflow:

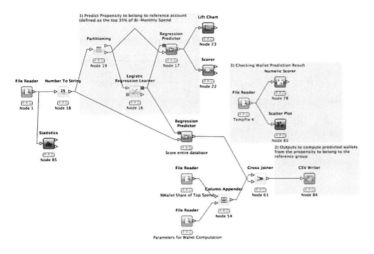

Figure 8-2 Wallet prediction workflow

- **High-spend reference account prediction**—By defining the highest-spending shoppers as the reference account (the top 5 percent in this case), a logistic regression model (learner node) is trained on a 50 percent sample partitioned using a random seed of 9999999. The DV of the model is the probability of being a reference shopper (that is, a high-value shopper). Lift and gain charts are used to ascertain how well the model can predict which of the shoppers are the reference customers. The results as shown in Figure 8-3 indicate that the top 10 percent of the customers with the highest propensities can capture about 60 percent of the reference accounts (a lift of 5.966). The errors of the model predictions are at 94 percent. However, in Table 8-1, all the reference accounts are classified as non-reference accounts. By lowering the threshold probability to 0.25, the model as shown in Table 8-2 can then identify 6 of the 13 reference accounts and improve the accuracy to 97 percent.

- **Linking the propensity score to size of wallet**—It is assumed that the wallets of an account correlate to the strength of the propensity to being the reference account. The wallet share of the reference accounts is also computed from the distribution.

Based on these two assumptions, the propensity scores are converted to the wallets (details have been deliberately omitted because of the proprietary nature of the technique). Anyone interested may contact the author for assistance in the customization for his/her particular industry and products.

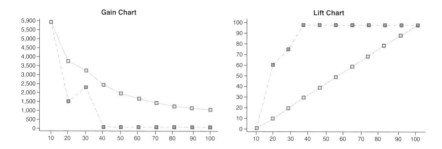

Figure 8-3   Reference Accounts Prediction Gain & Lift Charts

Table 8-1   Accuracy of Model Prediction

Accuracy = 94%	Predicted Non-Ref AC	Predicted Ref AC
Actual non-ref AC	210	0
Actual ref AC	13	0

Table 8-2   Accuracy of Model Prediction with Lower Threshold

Accuracy = 97%, Threshold P>0.25	Predicted Ref AC	Predicted Non-Ref AC
Actual ref AC	210	0
Actual non-ref AC	7	6

- **Checking of the BDC model performance**—By checking against the actual wallets, it was found the comparison between the actual and predicted wallets as shown in Figure 8-4 have a high R^2 of 0.829 and an RMSD of $452, which is quite acceptable when the largest spend was $4,084.

  The accuracy is further evidenced in Figure 8-5, which shows the limited scatter between the predicted and actual wallets. The accuracy is also higher for the higher-wallet customers, which is more important than predicting the lower-wallet customers.

Table "Scores" - Rows: 5	Spec - Column: 1
Row ID	**D predicted wallet**
R^2	0.829
mean absolute error	367.558
mean squared error	204,664.487
root mean squared deviation	452.399
mean signed difference	289.056

Figure 8-4 BDC wallet prediction accuracy

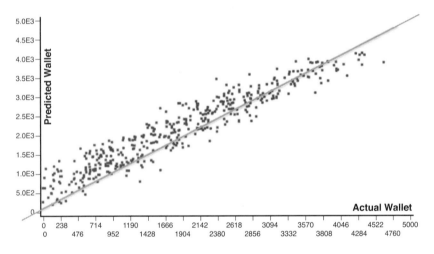

Figure 8-5 Comparison between actual and predicted wallets

Strategic Implementation

After determining the actual wallets, the following strategic actions may be implemented:

- **Effective strategies**—Used in conjunction with the propensities to purchase, the customers may then be divided into tiers: those who have large wallets, low penetrations, and high propensities to buy (large low-hanging fruits) may receive the highest-priority treatment. They are followed by customers with large wallets, low penetrations, and medium propensities (large fruits farther up the tree). For the customers with large wallets, low penetrations, but low propensities to buy (large fruits beyond reach), a new ladder is required to find out why such customers are not buying and to find ways to entice them to buy. Because the individual customers are known, you can then easily test the different strategies to fix either the pricing or values to increase wallet shares among this group of under-served high value customers.

- **Optimize return on investment (ROI)**—By testing and finding ways to minimize costs and increase conversions among the various tiers of customers, some customers may need to be fired for having small wallets and are not worth pursuing, whereas others with large wallets and low shares may not be worth the high costs to pursue them, also indicating that the fundamental value proposition and focus may not appeal to this group of customers. The senior management may need to invest more to acquire this competency and increase their values perceived by these customers.

- **Values, sentiment, and persona**—The failure of many campaigns often attributed to analytics were actually caused by the mismatch between marketing messages, offers, and target customers. Marketing materials and offers were usually worked out before the analytics teams were engaged to prioritize contact cadence and media choice. A better way is to create an intersection between the marketing team responsible for the contact collaterals and the analytics team, as laid out in Chapter 5.

When customers are identified, the analytics team provides its detailed profiling from all available data to the media team for alignment of the marketing materials with the target customers. Surveying a select sample from these customers may help to answer any further attitudinal and motivational questions. The survey results should also be linked back to the overall customers via link analytics. Once the questions are properly answered via analytics profiles and the voice of the customer surveys, the changes can then be tested and optimized via design of experiments (DOE).

Predicting Individual Customer Behaviors in Real-Time

The challenges of many physical and online stores include how to predict certain customers' impending behavior and their values to the store the minute they "walk in" to the physical or online store. To do this well, there are a few prerequisites: identify, profile, know the values, and have validated strategies per group of customers. The various cases are discussed briefly here; you are encouraged to browse and dive deeper into the respective topics.

- **Identify the individual customers**—Not being able to identify individual customers in physical stores results in treating each customer the same way. However, there are new analytics and technologies today that may begin to solve this challenge. Let's examine them in three settings:

  - **Physical stores**—Stores have used radio-frequency identification (RFID), Bluetooth tags, cameras with facial recognition, and iris sensors to attempt to identify customers when they enter into the store. Not only is the technology far from mature, but such measures also tend to invoke an image of Big Brother. To avoid adverse publicity, stores tend to shy away from such devices. However, given the recent surge in the number of mobile smartphone users and the ubiquity of apps, real-time analytics can use location-based data to score, target, and serve up compelling offers and relevant coupons the minute customers enter the store. The analytics should happen in real-time (in seconds rather than minutes). To do this well, you need to ensure that the analytics scoring and targeting models are constantly updated and embedded within the database (converting into PMML or any other in-database SQL codes) for enhanced performance. Similar to online stores, DOE tests should be done to optimize the lever settings that produce the optimal effects.

- **Online stores**—The most common way to identify visitors is to use cookies, which are simple codes placed within the online visitors' browsers or on their hard drives when they come to your site. With some ad networks, you may employ third-party cookies to track visitors to other sites. So when particular visitors come to your site, you not only know whether they are returning visitors but also the sites (that are also members of the same ad network) that they have visited. With cookies, you may only know the sessions belonging to the same visitor without knowing the true identity of the visitor. To know this, you need visitors to register and log in to their accounts.

- **Call centers**—Speech analytics may be used to detect the identities of the callers and also their sentiments. After a short conversation, the analytics should also be able to identify the particular topics the callers are calling about. An automated call routing would then route the callers to the most experienced customer relationship management (CRM) reps who are best equipped to handle these types of calls. Because speech analytics can identify key words and phrases within hundreds of thousands of conversations, it can also help determine the performances of customer service agents and whether the problems were satisfactory resolved.

- **Profile the individual customer**—Although most businesses would use the rich information collected on the online or mobile visitors for ad placement, few actually profile the visitors to do better customer segmentation, deliver more relevant contents, improve customer experience, and optimize values to customers and ROIs. The use of ad networks is a temporary fix because they do not have the end conversion data. A better way is to use the rich website contents to profile the online customers as they traverse the Internet. I usually use the analogy of identifying swarms of bees by the types of flower pollen they carry on their legs. Likewise, you can characterize the visitors to your site by the sites they previously visited. Besides profiling the visitors, these sites also need to be profiled. The attributes of the sites visited need to be characterized using text analytics and Natural Language Processing (NLP). The success of the analytics may be validated by the impacts the attributes can contribute to the enhancement of the model lifts for improving the customer experience, conversions, and value perceptions.

- **Knowing the visitor value**—Regardless of whether visitors are coming to your physical or virtual stores, once they are identified in real-time, you can very quickly use the last example to predict their wallets and LTV per category or per store. From this information, you can then decide how much you should invest in providing different groups of customers with the type of treatment their values deserve. The current one-size-fits-all approach and deluge of poorly targeted ads create poor customer experience and result in declining customer satisfaction and loyalty.

- **Proactive analytics-driven, customer-centric strategies**— To summarize, once you have performed these three steps, you can then do the following to develop analytics-driven, customer-centric strategies:

  - **Score the entire customer database**—Use predicted wallets, LTVs, propensities to buy more per category or overall;

then combine with existing demographic, behavior (online and mobile), and transaction data.

- **Launch a BAP initiative to develop effective strategies**—Set appropriate business goals for revenue growth, cost reductions, margin increases, developing new competencies, transformation, or turn-around; then assemble the sandbox infrastructure and team.

- **Strategic formulation**—Based on the specific goals chosen, all the analytics tools and methodology may be used creatively to help the team innovate new solutions. In general, I suggest the following:

  - Set up a 2x2 matrix of wallets and current spends, and determine (based on your goals) which of the four quadrants you want to focus on. For example, for revenue growth, you may want to focus on customers with predicted high wallets and high-to-moderate spend because they are lower-hanging fruits. For lowering costs, you may want to focus on moving the promotional and other expenses from the low-wallet and low-spend customers to those with higher wallets and higher likelihood to buy. For defending your base, you may want to focus on the high-wallet and high-spend customers to protect them from potential attrition that may occur because of market or competitor pressure.

  - Set up a 2x2 matrix of wallets and propensity to buy in the current purchase cycle. In this view, shown in Figure 8-6, the following were predicted:

    - The customers with high wallets, a high propensity to buy, and who buy a lot are in Region A. These are your most valuable and loyal base of customers and should be treated as such. They should be defended and monitored for any sign of attrition. If detected, effective counter measures must be set up and tested.

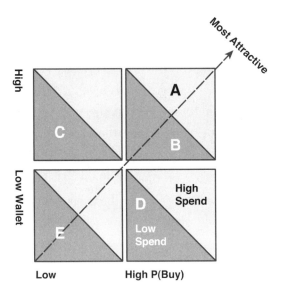

Figure 8-6 BDC wallet analytics strategy

- The customers with high wallets, a high propensity to buy, and yet buy little are in Region B. There might be gaps in the promotion or coverage (that is, neglected customers). Because these customers have been identified, efforts should be made to interview them and review the entire contact flow to ensure that the gaps are bridged and the appropriate wallet shares are attained as predicted.

- The customers with high wallets, a low propensity to buy, and who buy little are in Region C. They need special attention because promotional efforts and product values to date have failed to entice them to buy more. The traditional market research surveys and focus groups should be directed to this specific group to discover why they are not buying more and what improvements are needed to excite them so they want to enter a deeper relationship with you. Again, rather than taking these research insights and simply applying them, a series of test-and-learn DOEs should be set up to optimize the lever settings as informed by the research finding. They may simply involve changes to marketing messages, media mix, pricing, and packaging. Or, in severe cases, successful activation of this group of customers may require a wholesale transformation of the business, which may prove to be either too expensive or too hard to execute given the existing business realities. This is a decision not to be taken lightly that can be made only by the CEO with the support of the board.

- The customers with low or medium wallets; a high propensity to buy, and yet buy little are in Region D. These customers are low-hanging fruits that may again have been neglected. Tests of lower-cost options should be used to induce them to buy more. Obvious gaps in marketing and product promotions should be closed with measures that have been validated to be working.

- The customers with low wallets, with a low propensity to buy, and who buy little are in Region E. They should be excluded from any marketing or promotional plan. Although exclusion is one of the hardest decisions for a business, it is essential. Without excluding this group, focus would be diffused and precious capital diverted from serving the higher-value customers. Again, the effects of excluding such customers should be included in the overall test-and-learn DOEs.

Lever Settings and Causality in Business

In the early days of analytics, one of the most common applications was in targeting those who were most likely to convert. In terms of targeting, as long as you can differentiate between those who will and will not respond and convert, causality is not important. The problem with this approach is that it is clearly an example of diminishing returns with time. Those who are amenable to the business-as-usual marketing and product promotions are converted early and contacted often. Those who are harder to convert remain elusive. This is one of the reasons why many analytics projects seem promising at first, but begin to lose steam after a few campaigns. Unfortunately, most businesses that employ such targeting analytics approaches invariably end up spamming their most valuable customers.

This is how it usually happens. When the conversion rate starts to stagnate and decline, the business usually goes deeper into the list by emailing those with a lower propensity to convert. This drives up the costs (even though email costs are negligible, the damages to the customer relationship are real losses of revenues) and also further depresses the conversion rate. The only lever left is to email more to the high-propensity targets because it is the cheapest option.

This scenario happened in one of my prior engagements, when the effects of mailing started to wane because most customers were already inundated with too much mail and email per week. The senior management decided to go big by mailing even more in the hope of propping up the sagging results. This worked for a while, until the customers started to get annoyed and opted out *en masse*. Clearly, this is not the best way to treat your best and most loyal customers by spamming them.

Fortunately, there is another way, and someone with a more mature analytics competency will be able to do it differently. Rather than simply "targeting" the customers, the BAP approach can be adopted to study *why the customers are not converting*. From segmentation, profiling, and wallet and LTV prediction results, hypotheses can be drawn up and potential levers defined. Control experiments on the various levers can then be set up to test the hypotheses. These levers may go beyond marketing and sales options; they may include new product features, perceived brand values, market sentiment, and competitive pressure. The controlled tests allow the business to know the incremental effects of the various levers, optimize, and size the opportunity and ROI before full scale roll-out.

Start with a High Baseline

Before I get into the details of control groups in multivariate tests, I want to caution you when you start this for the first time. Do not treat this as an academic exercise—you do not need to test every single combination of factors. Treat it as a business exercise to win. Start with your best learning to date and build a high baseline. Once you established the baseline, test factors will be able to elevate the results further from the baseline. Any factors that end up depressing the baseline should be discarded. This way, you can conduct the test-and-learn exercise without sacrificing your business results!

Causality with Control Groups

Let's work through a case to contrast the differences between the conventional AB tests with the fractional factorial (multivariate, multicell tests). Assume a case of testing the incremental effects of three factors (offers, messages, and email reminders), each with 2 levels (options 1 and 2) and the results of response rates. The factors may also be other channels, discounts, coupons, fonts, colors, and design copies (and can have more than two levels). A full factorial combination of the 3-factor, 2-level tests give you 2x2x2 or 8 tests, as shown in Table 8-3.

Table 8-3   Design of Experiment (Adapted from the *National Institute of Standards and Technology [NIST] Statistical Handbook*[2])

Test	Offer	Message	Email	Results
1	1	1	1	3.3%
2	2	1	1	6.3%
3	1	2	1	4.1%
4	2	2	1	5.7%
5	1	1	2	5.7%

---

2 http://www.itl.nist.gov/div898/handbook/

Test	Offer	Message	Email	Results
6	2	1	2	5.1%
7	1	2	2	5.9%
8	2	2	2	5.3%

Let's use the case adapted from NIST and assume that eight tests were carried out for using different options and their subsequent response rates. To isolate the effects of each of the factors, let's take the average of the results with the particular factor. For example, to get the effect of offer 1, take the averages of all the results with offer 1, or (3.3 percent+4.1 percent+5.7 percent+5.9 percent)/4 or 4.75 percent. You can also obtain the effect of message 1 and email reminder 1 of 5.1 percent and 4.85 percent, respectively. You can similarly obtain all the other main effects and the secondary effects (with two or three factors). In this case, 2^3 or 8 tests are not too many. However, in real situations, there are usually more than ten factors, each with five or more levels. The number of tests would therefore be 5^{10}, which is a very large number (~10 million trials) that would take a while to test. Assuming each trial takes only a minute, it will take 18.6 years to complete!

To reduce the number of tests, you can use a half factorial design (using half [four] of the eight full factorial tests) or other fractional factorial designs, which need to satisfy certain conditions. Because this is beyond the scope of this book, it is not covered here. (If you are interested in more information, browse the Internet or refer to the *NIST Statistical Handbook*.)

There are many options for the half factorial design. One is the use of just the 1, 4, 6, and 7 tests. The effect of offer 1 is now 4.6 percent (cf. 4.75 percent), message 1 is 4.2 percent (5.1 percent), and email reminder 1 is 4.5 percent (4.85 percent). Though the numbers have changed, the number of tests is reduced by half. Tools such as SAS, JMP, and Minitab help you design the experiments and compute the effects of all the different options.

Conclusion

Similar to the intersections that engender innovations to flourish from the so-called Medici effect, the apparent unknowable events may present unique opportunities in business. Analytics deciders likewise welcome such challenges because they thrive within the intersection of ideas.

This chapter discussed four seemingly unknowable (UU) events and how to know the unknowable. One challenge arises from missing or dirty data. Another issue is incomplete views. The third is the difficulty posed by the need to predict individual customers' behavior in real-time. The last is the difficulty of isolating the incremental effects of factors and their true causal relationship in affecting outcomes, which optimizes the factors to produce the best outcomes.

Besides these examples, there are many UU events that await the creative treatments of analytics deciders in transforming them into lucrative business opportunities and success over their competitors.

9

Demonstration of Business Analytics Workflows: Analytics Enterprise

"If you don't know where you are going, you'll end up somewhere else."

—Yogi Berra

By the time you reach this chapter, you have hopefully gained some insights on where you want to be in terms of the analytics journey. Just a recap: By now, you have looked at the data and how it is managed; been introduced to business intelligence (BI) and advanced analytics tools; learned the importance of the business analytics (BA) process; saw how analytics can transform enterprise business processes; and learned how to apply analytics to recognize patterns in business and to know the unknowable.

You now look at how analytics can be leveraged to solve sample business problems. In this and the next chapter, the top ten questions discussed in Chapter 1 are addressed, leveraging everything covered so far. The focus is not on coding or building analytics models, but on how the problems may be solved using the business analytics process (BAP) and its standard components.

A Case for Illustration

- **Background**—Kathy, the new CEO of a mid-sized grocery chain, came upon a recent news article on how Kroger has successfully leveraged analytics[1] and grew its sales by 6.6 percent in 2012. This is more than twice the sales growth of its next five leading supermarket competitors! She decided to gather her team and see what they could do to replicate Kroger's success.

- **Critical questions**—To get the team prepared for the meetings, she presented team members with a list of ten questions.

  - **Describe current conditions:**

    1. How is our business doing?

    2. Where have we been investing our marketing dollars by channel and regions?

    3. How are the revenues and profits by stores, regions, and products/categories?

  - **Diagnose where the challenges lie:**

    4. Where did we do well and where did we do poorly?

    5. Were the poor performances recent occurrences? What was done to remedy the situations and what were the results?

  - **Predict outcomes and test potential levers:**

    6. Given everything we know, could we have predicted this year's results from last year's data?

    7. If so, how can we base on the predictions and test the effective levers so we could have done differently? What confidence do we have in the business outcomes before major investments are made?

---

1 http://adage.com/article/special-report-marketer-alist-2013/customer-data-pushes-laggard-kroger-digital-forefront/243803/

8. Taking the validated predictions, can we confidently predict next year's results with this year's data?

• **Prescribe what should be done for optimum results:**

9. Given the reduced marketing budget for next year, what options do we have?

10. If we were to increase our marketing budget by 10 percent, how should we spend it and how would it affect our results for next year?

Knowing the different types of solutions she is looking for, Kathy divided her questions into four sections (see Table 9-1). The first five questions look at what happened to date (descriptive) and pinpoint the trouble areas (diagnostic). The last five questions look for future options (predictive) and through analytics to obtain the optimum solution (prescriptive).

The questions should be answered sequentially, and the latter questions depend on how the preceding questions are answered. For example, without a thorough understanding of the current situation, attempts to answer Q4 and Q5 may lead to wrong diagnostics. Similarly, without Q6–Q8, decisions would have to be made based on past results and experiences.

Table 9-1 Overview of Business Analytics Solutions

Types of Analytics	Descriptive	Diagnostic	Predictive	Prescriptive
Question	Q1–Q3	Q4–Q5	Q6–Q8	Q9–Q10
Tools	BI and visualization.	Advanced BI visualization and research.	Advanced predictive analytics models.	Advanced optimization and decision management.

Types of Analytics	Descriptive	Diagnostic	Predictive	Prescriptive
Possible Insights	Most revenues from shoppers spending between $400–$500 per month. A large number of low spenders are not profitable. Social media promotion is costly but shows better results than flyers.	Doing well in packaged goods but poorly in fresh and organic produce. Some low spenders tend to shop occasionally, but would pick up organic and fresh produce on promotion when buying milk or household items.	Found low spenders that are affluent shoppers with predicted grocery wallets >$250/week who live near the stores but rarely shop at the stores. Predicted 1,000 customers per store who would likely increase spending by >$100 per week. They would shop across categories if given a series of mobile coupons along with user reviews. The mobile offers appear when the shoppers enter the stores as part of their shopping planning app would result in a high redemption rate The program would double the margins from such high-value customers.	Based on tests, suggest the right mix of mobile ads and events to maximize marketing returns on investment (ROIs). Daily views of the actual results vs. pro forma projections and prescribe solutions and tests.
Business Value	Better grasp of what the shopper spends. Identify areas of high and low performance. Identify channel investments and performance.	Pinpoint where the weaknesses lie. Identify some potential opportunities.	Better knowledge of their true values and purchase preferences. Ability to precision target high-value customers and improve ROI. Improved assortments and margins by encouraging cross-category purchases.	Increase number of visits by 20 percent and size of basket per shopper by 30 percent. Lower attritions by 10 percent.

For example, the question about how to deal with a drop in revenues for a supermarket chain may be broken down into the following subquestions:

- **Not where, but who?**—Instead of the usual way of isolating the revenue drops by stores, regions, and products, a more useful view is to ask who among the shoppers are buying less or (worse) have stopped coming into the store. Looking inward is intuitive and easier, and can tell you what went wrong and where. However, it rarely can tell you the reasons why specific customers are leaving you or, more importantly, give you ways to remedy the problem.

- **Customer-centric view?**—To properly answer this question, you need to know the shoppers intimately in terms of what they buy, why they buy, and how they buy. From these answers, various shoppers can then be grouped together into segments (which must not be inward-looking groupings based on products or operations, or even how much the customers buy).

Segments in this book are used to mean the distinct groups in which customers group themselves by their behavior, attitudes, needs, and wants from their underlying utilities. As a result, they may *respond similarly to an ad campaign and share similar brand perceptions.* This distinction is important because segments are commonly used to denote the groups of customers viewed by business from the angle of any prior knowledge. I find that such inward-looking segmentation often is not very actionable and may lead the business strategic planning down the wrong path.

The secret of successful marketers is therefore to identify such customer-centric segments and find ways to engage, excite, and satisfy the customers within the targeted segments. Such segments may behave differently over time, depending on the shifts in product trends, culture, and opinions of influence leaders. However, within the same segment, customers tend to think, feel, behave, and purchase similarly.

- **What and how do the segments buy?**—Once the segments are determined, the types of products they have been buying to date, and how they buy, can be profiled at the centers of the groups by taking the group means or medians of the various segment attributes. Done properly, these buyers should be quite similar to each other in terms of their purchase behavior and persona. Another indication that the segments are defined properly is that the specific segment to which any shopper belongs should remain pretty much the same over time.

- **What has changed?**—Although the segment membership may remain fairly static, the purchase behavior may change significantly due to the changes in market conditions, promotions, product features, and pricing. So businesses must always monitor the trends of how such customer segments interact with their products, promotions, and brands.

- **Why and how important?**—Understanding what has happened is just half the answer. You need to know the underlying reasons, evaluate their impacts on business, and be able to prioritize various intervention options or levers.

- **What can be done?**—Once the possible levers are identified, the incremental effects of different levers can then be isolated and validated with tests (with properly designed multicell control groups) that are commonly known as the design of experiments (DOE). CapitalOne made DOE popular in its direct-mail credit card promotional campaigns. Its stellar performance (with EPS and return of equity >20 percent[2] for each one of its first 10 years as a monoline credit card business) was attributed to CapitalOne's use of customer data—targeting analytics and tests using DOE for everything it does.

---

2 http://en.wikipedia.org/wiki/Nigel_Morris

- **What-if scenarios planning?**—Once the effects of the levers are determined, a what-if scenario simulator can then be built to optimize the combination of the most effective levers to produce the desirable business outcomes. Along the way, performance metrics can be used to update the simulator to ensure accuracy and reliability.

Figure 9-1 shows this workflow graphically.

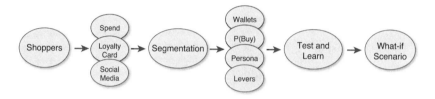

Figure 9-1   Supermarket shopper revenue analytics workflow

Following the schematics as shown in Figure 9-1, let's revisit the top ten questions alluded to in Chapter 1. I will cover all the questions, except the questions pertaining to customer relationship management (CRM): Q4–Q7, which are covered in the next chapter. I have expanded the questions into several subquestions, which are intermediate steps to answer the larger questions. You are encouraged to check those that you know how to apply analytics to solve and with which your companies are currently actively working on the solutions.

Top Questions for Analytics Applications

As an example, template is filled out for the first example. The empty templates below the questions are meant for you to see whether you can sketch out the BAP to answer the questions that are relevant to your business.

Financial Management

The following sections discuss the typical questions (and answers) business owners have about how to effectively maximize the management of finances and how to better plan their business operations, strategy, and direction.

Are Our Goals the Right Goals?

Based on data and analytics:

❏ What should our business goals be (revenue or margin growth, market share gain, or be a market leader)?

❏ What should be our quarterly and annual financial goals?

Based on Table 9-2 and as mentioned in Chapter 1, begin with the customers and predict their wallets and shares by segments. The wallets and shares of wallets will be predicted using loyalty cards (or any other means) to link the transactions and behaviors to individual customers. The historical data on past investments and changes in the customer segments in terms of revenues and wallets may be used to determine what the business goals should be. Using the demographics, segment persona and their respective P(Buy), their propensity to buy, coupled with the past performance of the lever settings (marketing, sales, services) and their respective investments, what-if scenarios may be built for different investment mixes. By varying the investments, the optimal amount and distributions may be obtained to achieve certain quarterly or annual financial goals.

What Areas of Opportunities?

Based on analytics, where are the areas of major revenue or profit growth opportunities (see Table 9-3)?

❏ Where are the areas of greatest opportunities for revenue growth?

❏ Where are the greatest cost savings and the most readily realizable?

Table 9-2   BAP Checklist—Financial Goals

Modules	Spend Data	Loyalty Card	Demographics	Company Data	Social Media	Segmentation	P(Respond)	P(Buy)
Needed	X	X	X	X		X		X

Modules	P(Churn)	Trends(Time Series)	Wallets	Persona	Levers	DOE	What-if Scenario	Scenario Optimization
Needed		X	X	X			X	X

Table 9-3   BAP Checklist—Opportunities

Modules	Spend Data	Loyalty Card	Demographics	Company Data	Social Media	Segmentation	P(Respond)	P(Buy)
Needed								

Modules	P(Churn)	Trends(Time Series)	Wallets	Persona	Levers	DOE	What-if Scenario	Scenario Optimization
Needed								

Right Investment Mix?

Based on analytics, is the distribution of our investments optimal (see Table 9-4)?

❑ Are our investments adequate and the mix optimal?

❑ What should we do to maximize impacts within the budget constraints?

Table 9-4   BAP Checklist—Investment Mix

Modules	Spend Data	Loyalty Card	Demographics	Company Data	Social Media	Segmentation	P(Respond)	P(Buy)
Needed								

Modules	P(Churn)	Trends(Time Series)	Wallets	Persona	Levers	DOE	What-if Scenario	Scenario Optimization
Needed								

Human Resources

Next, we look at management of the critical enterprise assets in human capital.

Are Analytics Talents Properly Managed and Organized?

(Instead of using the previous template, you might want to consult Chapter 11 before checking the questions in this section.)

Talent Management

Proper management of analytics talents includes the following:

❑ Do we have the right talents to enable the analytics business transformation?

❑ Are the key roles for the analytics talents properly designed and clearly communicated?

❏ Do we have the right HR setup for managing the requisite qualifications, training, performance evaluation, and recognition for different analytics roles?

❏ Do we have the know-how to recognize and develop such talents?

❏ Do we know how to recruit, train, reward, and retain highly-sought-after analytics talents?

Organizational Issues

In my many years of leading analytics teams, the most frustrating issues are usually organizational and personnel, rarely tools. So it is important to ask these questions about your company:

❏ Do we imbue the organization with analytics thinking and focus?

❏ Do we produce more analytics deciders among the current and emerging leaders (those who can comfortably bridge the two worlds of business and analytics)?

❏ Do we develop the best organizational and reporting structure for analytics talents at our stage of analytics maturity?

❏ Do we bridge organizational silos to better leverage enterprise analytics?

Internal Operations

Analytics may be applied beyond marketing and sales; in fact, any part of the business with data can be enhanced with analytics.

Are Our Business Processes Driven with Insights from Predictive Customer Analytics?

Business processes are linked to customer insights such as demands, LTV, product preferences, pricing sensitivity, and so on.

❏ Do our business processes leverage analytics customer knowledge assets into an effective enterprise analytics value chain (see Table 9-5)?

Table 9-5   BAP Checklist—Analytics-Driven Business Process

Modules	Spend Data	Loyalty Card	Demographics	Company Data	Social Media	Segmentation	P(Respond)	P(Buy)
Needed								

Modules	P(Churn)	Trends(Time Series)	Wallets	Persona	Levers	DOE	What-if Scenario	Scenario Optimization
Needed								

Examples of Analytics-Enabled Processes

❏ Is our supply chain responsive to the predicted demands at various levels?

❏ Do we know how to reduce the costs to serve by applying predictive analytics to all our processes (see Table 9-6)?

Table 9-6   BAP Checklist—Process Examples

Modules	Spend Data	Loyalty Card	Demographics	Company Data	Social Media	Segmentation	P(Respond)	P(Buy)
Needed								

Modules	P(Churn)	Trends(Time Series)	Wallets	Persona	Levers	DOE	What-if Scenario	Scenario Optimization
Needed								

❑ Do we have a proactive analytics-driven strategy to respond to the changing characteristics and the deluge of Big Data on customers, products, or suppliers?

❑ Do we still rely on aggregated information, averages, and generalized models for supply chain management (SCM)?

Are Our Sales Deployment and Performance Fueled by Analytics Insights?

Management of face-to-face or call-center sales reps includes the following.

Optimized Leads and Cadence

❑ Are our sales efforts optimized via analytics to route the highest-quality leads to the right salespersons with the right cadence? See Table 9-7 and check those analytics that would help sales efforts and cadence.

Table 9-7   BAP Checklist—Sales Leads and Cadence

Modules	Spend Data	Loyalty Card	Demographics	Company Data	Social Media	Segmentation	P(Respond)	P(Buy)
Needed								

Modules	P(Churn)	Trends(Time Series)	Wallets	Persona	Levers	DOE	What-if Scenario	Scenario Optimization
Needed								

Analytics-Empowered Sales

❑ Do our sales reps have the necessary analytics customer insights to guide them to the right scripts and offers to maximize their chances of being better able to serve the customers with a certain segment persona (for example, customer preferences, value rankings, prior purchases, and prior service calls)?

Table 9-8   BAP Checklist—Analytics Insights for Sales

Modules	Spend Data	Loyalty Card	Demographics	Company Data	Social Media	Segmentation	P(Respond)	P(Buy)
Needed								

Modules	P(Churn)	Trends(Time Series)	Wallets	Persona	Levers	DOE	What-if Scenario	Scenario Optimization
Needed								

Salesforce Optimization

❑ Do we have the right number and proportion of reps to handle the different types of customers and reasons for contacts, both inbound and outbound?

Table 9-9   BAP Checklist—Salesforce Optimization

Modules Needed	Spend Data	Loyalty Card	Demographics	Company Data	Social Media	Segmentation	P(Respond)	P(Buy)

Modules Needed	P(Churn)	Trends(Time Series)	Wallets	Persona	Levers	DOE	What-if Scenario	Scenario Optimization

Sales Performance Analytics

❏ Are the sales performances and customer knowledge captured through DOE test-and-learn metrics shared?

Table 9-10   BAP Checklist—Sales Performance

Modules Needed	Spend Data	Loyalty Card	Demographics	Company Data	Social Media	Segmentation	P(Respond)	P(Buy)

Modules Needed	P(Churn)	Trends(Time Series)	Wallets	Persona	Levers	DOE	What-if Scenario	Scenario Optimization

Conclusion

This chapter laid out the ways in which analytics components can be combined to solve various business cases. You are encouraged to follow the logic behind various workflows and apply them to your own business problems and data. To aid you in the exercise, a graphical representation of the analytics workflow with a checklist of some of the important business applications were given. A template of analytics components was also provided that you can use to assess your company's current level of analytics applications and to practice answering the various questions with analytics components.

I hope that once you are proficient in breaking down business problems into their constituent questions, you can formulate smaller workflows in KNIME as metanodes and link various metanodes into a complete workflow to solve the intended problems.

10

Demonstration of Business Analytics Workflows—Analytics CRM

"Spend a lot of time talking to customers face to face. You'd be amazed how many companies don't listen to their customers."

—Ross Perot

Although savvy business owners know the importance of listening to their customers, the day when business owners could personally talk with each customer is long gone. "Talking" to customers today can take the form of surveys or focus groups, but such surveys may not be representative of the voices of all the customers, and they are costly and take time to conduct. Focus groups are even more expensive and their results are not easily quantifiable.

In 2001, I was a member of the IBM market intelligence (MI) senior management team. When presented with the findings of millions of dollars of market research results, I realized that much customer intelligence can be derived from the customers' behavior data, especially their transaction data. I commented to the VP of market research that "if instead of survey questions, we presented to our customers our products and solutions with their values, and if the customers answered the questions not with words, but with their purchases, we might be able to 'listen to the voice of the customers' through data analytics!" Today, with social, local, and mobile (SoLoMo) media data and other ubiquitous data sources, much customer knowledge can be

219

obtained in a timely and cost-effective manner as long as the business knows how to deploy analytics.

This chapter demonstrates how analytics can be leveraged to help businesses understand and better serve their customers.

Questions About Customers

First, you need to answer questions about how you can know the customers more intimately than your competitors. These questions include who the customers were before, are now, and will be in the future. How can you characterize them in terms of segments? And instead of numbers and facts, how can you know them as real people with their respective persona so you can treat them accordingly? Once you perfect what you do, you can determine ways to effectively engage customers by knowing to which segment they belong. However, when they first come on to your web sites or call, how do you know what segment they may belong to? Defining a few questions by the way customers would answer them can help analytics models predict their segment membership. Such questions, sometimes known as golden questions, have been used by sales teams to identify ways to successfully close the sale. You should check the following list against what you know and are doing when applying analytics to care for your customers in your business.

Know the Customers

Following is a list of likely questions business may ask about their customers:

- **Intimate customer knowledge**
  - ❏ Who are your customers, past, present, and future? This may be answered by a business intelligence (BI) analysis and

visualization of the customer transaction and loyalty card data for the past few years. As for the future, a series of time-dependent holdout models can be done to predict the next month's or quarter's spend of existing customers, and the number of new customers predicted to be added as well as revenues from such newly acquired customers.

❏ How many customer types are there in terms of how they buy, how they perceive the brand, and why they buy? Once grouped or segmented analytically, the following questions on customers must be answered for the respective segments beyond just for generic customers:

    ❏ What is the respective segment persona? A segment persona may be created based on surveys, augmented demographic data, and segment behavior highlights. From the segment persona, media planners who know the customers in a particular market can formulate the best strategies and tactics to excite and satisfy the customers in the particular segment.

    ❏ Do customers within each segment show a similar persona, product preferences, and responses to marketing and sales campaigns? Once a segment persona is defined, marketing and sales levers have to be tested and optimized. These levers can include different offers, events, messages, creative, product features, after-sale service, pricing, media, social groups, and key influencers.

    ❏ Do we have the two or three questions to ask any visitors to your physical or virtual stores, to determine which customer segments they belong to? These are the "golden questions" that determine which segment customers or prospects are most likely to belong to. They can be asked in person, over the phone, or by a browsing pathway through cleverly designed web or mobile sites. Once this is determined, the right sales or marketing collaterals that were validated to be effective can then be used to boost the probability of successfully engaging them.

Table 10-1 shows a customer analytics template. The checked modules are again those modules that are essential to an analytics workflow to answer the questions above for knowing the customers better.

Table 10-1   Customer Analytics Template

Modules	Spend Data	Loyalty Card	Demo-graphics	Company Data	Social Media	Segmentation	P(Belong to Segment)	P(Buy)
Needed	X	X	X	X		X	X	

Modules	P(Churn)	Trends(Time Series)	Wallets	Persona	Levers	DOE	What-if Scenario	Scenario Optimization
Needed		X	X	X	X			

Actionable Customer Insights

In addition to simply knowing the customers, following is a list of questions on actionable customer insights. The ability to reliably answer these questions and to obtain actionable insights will be critical to the success of any business. For each of the questions below, the reader is encouraged to use the templates in Table 10-2 to outline the workflow. For example, for the question on predicting customer's propensity to buy certain product, the workflow would be (Spend + Demographics Data) + (P(Buy) model) + (DoE).

Table 10-2   Practice Template for Predicting How Customers Buy

Modules Needed	Spend Data	Loyalty Card	Demo-graphics	Company Data	Social Media	Segmentation	P(Respond)	P(Buy)

Modules Needed	P(Churn)	Trends(Time Series)	Wallets	Persona	Levers	DoE	What-if Scenario	Scenario optimization

- **How do customers buy?** This is predictive behavior; whether they buy from you or not.
- ❏ Purchase amounts.
- ❏ Predicted propensities to buy per product.
- ❏ Predicted wallets per products.
- ❏ Predicted lifetime value (LTV). The LTV can be predicted using the same Buyer Differential Characteristics (BDC) analytics as described in Chapter 8 except substituting the transactions with the net present value (NPV) of all the revenue or profit streams through the customer's life cycle with the business.
- ❏ Predicted media preferences throughout their purchase cycle. Predict their P(Respond) as per medium based on the past responses as the DV, and all other purchase patterns, demographics, and segment data as the Independent Variables (IVs).
- ❏ Predicted pricing sensitivity. Predict propensities to buy, given a particular price point so it is computed as a conditional probability $P(B|A) = P(A\&B)/P(A)$, where B is the product purchase and A is the particular price point.

❏ Predicted product and offer preferences. Predict propensities of a customer in purchasing a product when presented with an offer. The top preferences can then be obtained by choosing those products and offers with the highest propensities.

• **Customer loyalty**

❏ Are the customers satisfied with your products, services, and the ways you engage them? Although this may need to be combined with customer satisfaction surveys, changes in customers' propensity to churn over time may also be indicative of the customer loyalty. When a customer's propensity to churn decreases, it may indicate the customer's increased loyalty and vice versa. (Use Table 10-3 for practice in formulating the workflow in answering the questions.)

Table 10-3 Practice Template for Customer Loyalty Workflow Modules

Modules Needed	Spend Data	Loyalty Card	Demo graphics	Company Data	Social Media	Segmentation	P(Respond)	P(Buy)

Modules Needed	P(Churn)	Trends(Time Series)	Wallets	Persona	Levers	DoE	What-if Scenario	Scenario Optimization

❏ What are your values as perceived by your customers?

❏ How do you measure their loyalty status?

   ❏ Customer satisfaction (individual happiness).

   ❏ Propensity to churn within the next purchase cycle. Build a model to predict the propensity for any customer to churn in the next purchase cycle using past customer attrition data with time-dependent holdouts.

❏ Net Promoter Score (NPS).[1] It may be interesting to link customer loyalty measures such as a propensity for wallet share gain, a decrease in propensity to churn, and an increase in propensity to buy more from individual customers with the NPS surveys.

❏ Pricing sensitivity per customer loyalty per segment (cross-tab predicted price sensitivity with predicted customer loyalty per segment)

❏ Perceived brand equity per customer by loyalty and segment

- **New customer acquisition**

❏ What are the costs to acquire new customers from the respective segments?

❏ Based on the predicted LTV and segment persona, do you have the right value proposition to the target segment?

❏ Are you investing optimally in using the right message, offer, and media mix to acquire the target high-value customers?

❏ How can you reduce the acquisition costs while increasing return on investment (ROI)?

❏ Is the newfound acquisition strategy scalable to grow your base?

- **Churn intervention and win back**

❏ Do you know past and predicted customer attrition rates by segments and LTV?

❏ Do you know the probability to churn in the next week/month for every customer in your database?

---

1 Reichheld, Frederick F. *The Loyalty Effect*, Harvard Business School Press, 1996 (Revised 2001).

❏ What are the reasons why customers are leaving (by segment and value)?

❏ What can you do?

❏ How much should you invest to stop the attrition?

❏ Did you do scientific design of experiments (DOEs) to validate validity and incremental causal effects?

❏ How can you win back the high-value customers that may have stopped buying for a while?

• **Value-commensurate investments**

❏ Are you investing appropriately in terms of the customer segments' respective values and importance to the business?

• **Trending**

❏ Do you track and know how the analytics measures have been trending over time?

Social and Mobile CRM Issues

Given the growing importance of social and mobile channels in CRM, it is crucial that we pay attention to the right issues and applications. The following is a list of some important questions that BA may help to answer:

• **Are you focused on the right social and mobile issues?**

❏ **Issues**—Do you know the issues that customers and prospects are talking about on social media in real time?

❏ **Influencers**—Who are the main influencers?

❏ **Competitors' activities**—What are your competitors saying and doing in the social media? Do you have a consistent counterstrategy?

❏ **Sentiments and responses**—In case of emerging issues (positive or negative), do you have the strategy to learn, respond, and manage effectively?

❏ **Influencer engagement**—Do you have an ongoing strategy to engage and foster positive relationships with emerging social influencers?

❏ **Performance metrics**—Are there any performance metrics to monitor the effectiveness of your social and mobile media strategy?

CRM Knowledge Management

As more business analytics are applied to diverse data, it has generated much customer insight over time. It is more important now than ever to put in place a Knowledge Management system. Customer insight and knowledge are "perishable assets" that would "perish" over time unless properly managed. An effective Customer Knowledge Management system should permit the easy sharing, storage, testing, and application of customer knowledge as an enterprise asset. To do this, the following is a list of pertinent questions:

- Is knowledge management an enterprise asset? If you have checked many of the previous boxes, much knowledge has been generated. It is important to have a consistent knowledge management strategy, as outlined in the following questions:

❏ **Captured, codified, and managed**—Is the knowledge generated consistently captured, codified, and properly managed (searchable and reusable)?

❏ **Test and learn, DOE**—Are all insights scientifically tested before implementation? Ensure that the following boxes are all checked:

❏ Do you know the incremental effects of each measure and can you learn from them?

❏ Are properly designed, scientifically valid control groups used for all studies and activities?

❏ Are the causal effects and sensitivity of the various lever options properly established?

❏ Is the scientific DOE used?

❏ **Enterprise assets**—Is the knowledge readily accessible as an enterprise-wide asset?

❏ **Competitive edge**—Can the entire company leverage such knowledge as a consistent competitive edge?

Conclusion

This chapter continues to answer the remainder of the top ten questions for which analytics may be leveraged to provide answers and solutions. The focus is on the questions surrounding the most important asset of all businesses: value to their customers. One of the greatest challenges faced by today's businesses is the absence of in-person contacts with customers. As more and more businesses are conducted remotely or virtually, vital pieces of customer voice and knowledge are inevitably lost. This chapter began with an attempt to capture intimate customer knowledge and the voice of the customer from the use of analytics on customer data both big and small.

First, the business must know intimately who their customers are in terms of their persona and preferences. How can you identify the customers the minute they step into your store (physical or virtual, online or mobile)? To be successful, businesses must also engage customers using their preferred media and must be agile to change with changing mores.

Second, to differentiate high-spend and low-spend customers, analytics must be used to predict how much they might spend on your products, even if they buy very little from you today. One of the largest drains on valuable business resources is not knowing or prioritizing treatments based on customer values. Another is the lack of knowledge about customers' product preferences and how much they are willing to pay.

After a business can accurately predict individual customers' value and preferences, it can better serve its overall customers by delivering the right product at the right price through the right channel to respective customers who would value such service. Such customers tend to be the most loyal customers and spend the most. Over time, for such unrivaled service, they allow the business to earn larger revenues with greater margins than its competitors.

Finally, analytics can also help businesses acquire more high-value and potentially loyal new customers, and to retain more of them. If there are breakdowns in the service of such customers, they can be quickly detected and remedied by predicting impending churn. When such knowledge is generated through analytics, it can then be scientifically validated, captured, and added as an enterprise asset to the company's arsenal of competitiveness.

11

Analytics Competencies and Ecosystem

"If a leader demonstrates competency, genuine concern for others, and admirable character, people will follow."

—T. Richard Chase

To establish an analytics function, there are several prerequisites. Apart from the obvious requirements for rich and reliable data and analytics tools (discussed in Chapters 3 and 4), some softer requirements are also important. Just as you can have the best ingredients and kitchen gadgets, there will be no restaurant without the right chef and a smoothly running kitchen. The critical factors for the analytics team to function properly and be successful are the analytics competencies and a supportive ecosystem.

We can view this in terms of the analytics maturity model, as illustrated in Table 11-1. The five levels of maturity were slightly modified from the competency levels of Davenport *et al.* (2010).

Table 11-1   Analytics Maturity

Maturity	Level 1 Analytically Impaired	Level 2 Analytically Aware	Level 3 Analytically Inspired	Level 4 Analytically Proficient	Level 5 Analytical Leaders
Data access	Inconsistent, poor quality, poorly organized	Data in silos rarely shared or integrated	Data shared, but lacking central repository	Integrated, accurate, and analytics DB and DW	Relentless search for new data sources
Enterprise focus	None	Islands of data, technology, and analytics expertise	Early stage of an enterprise-wide approach	Key data, technology, and analytics resources centralized	Fully functioning BAP and well-established intersections
Analytical leadership	No awareness or interest, opinion-based deciders	Only at the function or process level	Leadership at select analytics decisions	Leadership fully invests in analytics competency	Strong leadership passion for analytics
Strategic targets	None	Tactically focused analytics	Limited strategic targets for analytics	Analytics activity centered on a few key areas	Transformation and competition-focused analytics
Analytics talents	Few skills in scattered talents	Isolated pockets of analysts and no analytics decider	Analysts in key target areas with a few analytics deciders	Highly skilled analysts in coordinated teams led by analytics deciders	World-class analytics deciders in CoE (Center of Excellence) and business units performing critical roles

The analytics maturity of any company can be measured along five dimensions, first proposed by Davenport *et al.* (2010)[1] as the

---

1 http://www.amazon.com/Analytics-Work-Smarter-Decisions-Results/dp/ 1422177696/ ref=sr_1_fkmr2_2?ie=UTF8&qid=1408848806&sr=8-2-fkmr2& keywords=davenport+analytics+competency

DELTA model. The acronym DELTA stands for five critical indicators of analytics competency (this case focuses on the business' analytics maturity instead):

- **D**ata access—Does the entire company have ready access to high-quality, rich, and integrated data?
- **E**nterprise focus—Do analytics resources have an enterprise focus or silo-focused?
- **L**eadership—Does the entire enterprise exhibit analytics leadership (that is, as analytics deciders, starting from the CEO)?
- **T**arget—Is analytics applied to strategic rather than tactical objectives?
- **A**nalytics talents—Does the company have enough analytics talents (both internal and external) from consultants and the analytics ecosystem)?

Analytics Maturity Levels

There are five levels of maturity: Level 1 is the lowest, indicating that a company is analytically impaired; most companies started from this default state. Level 2 is a company that is analytically aware. In this phase, companies start to realize the importance of data and analytics and start to do them in some parts of the organization, usually in direct marketing campaigns. In Level 3, the company is analytically inspired, where analytics has demonstrated some successes and started to be integrated into some of the company's strategies. Most of the more analytically matured companies today are in this stage.

The real differentiator comes in the next stage, Level 4. Companies that reach this level of maturity are recognized as analytically proficient. Despite many claims of maturity, few companies have attained this level of maturity. More and more of the top competitors in any industries are those that are on their way to reach the ultimate maturity in Level 5—as an analytical leader. The levels of maturity may also be indicated by how many business operations are done with fully functioning business analytics process (BAP) and led by analytics deciders.

The aim of this chapter is to show how to elevate the analytics maturity of business by addressing the important aspects of the DELTA model. (All issues related to data are addressed in Chapters 2 and 3. The remaining four factors are addressed in this chapter.) To be more mature along these four factors involves three critical issues. 1) *Organization structure*: What is the optimum structure for analytics based on industry best practices? 2) *People roles*: How should the people be organized and managed? 3) *Process*: What should the analytics process be to ensure continued success and the controls to know when it is not performing?

Analytics Organizational Structure

Over the last decade, I have witnessed analytically aspiring companies trying many different organizational models and reporting structures (see Table 11-2). There are three key dimensions in which the different analytics structures may be differentiated:

- **Reporting chain and accountability**—The members performing various roles in the analytics team need to be effectively managed and led. This dimension encompasses talent recruitment and training, skill development, retention, rewards, career progression, performance evaluation, chain of reports, and accountability. Analytics may exist as another silo and directly report to the CEO/president. It can also report into one of the other silos such as marketing, finance, business units, or IT. Each carries benefits and shortcomings.

- **Funding and sponsorship**—The size of the analytics function and hence how much funds to be allocated should depend on its success in supporting business. The best-case scenarios ultimately are for analytics to own its own profit and loss (P&L). However, it is not feasible in most cases in which the direct revenue contributions of analytics are not quantifiable. In any case, some level of investments needs to be allocated in the initial phase of the analytics launch. The funding may come from corporate; business units; or the function that analytics team is to report into marketing, IT, or finance. The analytics funding is simply part of the overall budget for the unit.

- **Knowledge repository and sharing**—Given the need for continuous innovations in analytics, there has to be robust knowledge management in place. Knowledge management comprises knowledge creation, knowledge community, knowledge capture and sharing, and knowledge recognition.

Table 11-2   Analytics Team Structure

	Reporting and Accountability	Funding and Sponsorship	Knowledge Management
Centralized team	Corporate	Corporate	Centralized
Consulting team	Independent	Corporate	Limited
Decentralized team	Business units	Business units	Decentralized
Center of excellence	Corporate	Business units	Centralized

Based on the preceding criteria, there are four options to structure the analytics function: centralized, consulting, decentralized, and center of excellence models. I will lay out the various models along the three dimensions and also their respective advantages and risks.

Let's look at how these structures differ (see Table 11-3).

Table 11-3   Summary of Analytics Organizations

	Benefits	Risks and Shortcomings
Centralized team	Clarity, synergy, career options	Ivory tower, irrelevance
Consulting team	Responsive to business	Difficult to launch and make a difference
Decentralized team	Responsive to business	Redundancy, silos, lacks strategic focus, talent attrition
Center of excellence (CoE)	Responsive to business, knowledge sharing, synergy, career options	Split ownership, confusion of allegiance, irrelevance, ivory tower, and decentralized

The Centralized Model

The *centralized* analytics team is relatively easy to set up and manage, so it is the preferred structure for most analytics startups. Here are the pros and cons:

- **Pro**—It has the added advantages of being *synergistic*, and if run properly, can result in:
  - **A knowledge-sharing** learning community.
  - **Job satisfaction**. By consolidating into one team, it may also provide sufficient bench depth for career development and progress. This is important because the lack of career advancement is often quoted as the major reason for talent attrition.
- **Con**—The team is taking directions and funding from the corporate office, which often tends to be functionally and geographically removed from the actual business in the field. Over time, the centralized analytics team can take on projects that may be strategic in nature, but of little relevance to the business units in the field. Because the P&L responsibilities often are in

the hands of the business units, few incentives are felt by business units to fund projects that have little impact on their P&L. Even if the corporate office levies a poll tax, the result would be the creation of a centralized analytics ivory tower. The centralized team may be very busy and productive, but little of what it does gets implemented and in the end have any impacts on business outcomes. To avoid this from happening, the centralized model adopted in the beginning must evolve quickly over time to one of the other models listed here.

The Consulting Model

The *consulting* model is the opposite of the centralized team in terms of reporting and funding structures. Instead of being tightly managed and funded by the corporate office, it is left to fend for itself. It is even encouraged to take on projects sometimes even outside the company. Dunnhummby of UK's supermarket chain Tesco is a successful example of the consulting model. Here are the pros and cons of this model:

- **Pro**—The advantages of the consulting model is that it encourages the team to be innovative. Facing varied projects and analytics challenges, a consulting team would need to continue to innovate and find new business. Instead of being an order taker, the team can essentially pick and choose projects they determine to be worthwhile based on their impacts on business and then charge commensurate rates. Such budgetary freedom and interesting projects means that the consulting team can afford to hire and retain top talents. The fact that Dunnhummby has become one of the leading analytics consulting teams for retail industries globally is clear evidence of the success of this model.

- **Con**—On the flip side of its success, the consulting model can be expensive, take time to set up, and may not always succeed. Due to legacy practices in which analytics is a novel entity, the existing teams may feel threatened, especially if they are already quite successful. Therefore, it may be hard to gain traction for an internal consulting analytics team.

  Even if the team is successful, the problem of few impacts on the founding company may persist. The team may simply be more successful outside of the company than within. For example, Dunnhummby was credited for helping Kroger attain faster revenue growth than Walmart and its closest competitors,[2] and yet could not stop Tesco's recent sharp decline in sales (reported to be the worst in two decades).[3]

The Decentralized Model

The *decentralized* team is the *de facto* structure when business or functional units start to embrace analytics by hiring their own analytics teams. Each unit specifies and funds the kind of analytics team that is most relevant to its line of business:

- **Pro**—Due to their respective business focus, decentralized teams are usually agile and responsive to business needs. Their size and scope are dependent on the teams' success in supporting the respective business units. Because their success is tied to actual business results, they tend to be relatively free from corporate politics and rivalries for funding that is common among functional units that do not have P&L responsibilities. As a result, the decentralized analytics model is easier to set up and manage over time.

---

2 http://adage.com/article/special-report-marketer-alist-2013/customer-data-pushes-laggard-kroger-digital-forefront/243803/

3 http://www.bloomberg.com/news/2014-07-29/tesco-had-worst-sales-decline-in-two-decades-kantar-data-shows.html

- **Con**—Because of its limited scope, the decentralized team might end up doing repetitive and redundant work from the enterprise perspective. Because each business unit must have its own analytics talent for support, there are often issues such as an overlapping of talents among business units, short career ladders, and lack of intellectually challenging projects. Such an environment often leads to a low level of job satisfaction due to limited training, team synergy, and career progression. With the current high demands and frequent job changes of analytics talents, severe talent attrition could ensue. Those who remain on the decentralized team tend to be those who prefer the status quo and usually produce very few new ideas or innovations. The team lacks strategic focus because it is preoccupied with fulfilling tactical demands and therefore often has limited impacts to the enterprise's overall business.

The Center of Excellence Model

The *center of excellence (CoE)* model has become quite popular in recent years. The basic concept of the CoE is to consolidate the critical strategic analytics assets in a central team under an analytics leader. The CoE is not directly funded by business units, but as part of the corporate office. All R&D is centralized at the CoE. The alignments to business are ensured by analytics strategists embedded within different business units who focus on the application of analytics at the business units.

When I was with IBM in 2005, I was on the management team that was responsible to set up the database marketing and analytics CoEs within the corporate marketing function. Our tasks were to centralize the strategic assets and offshore the less-strategic functions. The CoE served as the caretaker of the enterprise analytics focus, while the less-important tasks were devolved to the respective supporting teams outside of the company. Here are the pros and cons:

- **Pro**—When working well, the CoE model provides synergy between the various business units and the corporate analytics CoE. Because parts of the CoE are the analytics strategists within each business unit, they are the links between the business and the CoE. When functioning properly, they serve as the bridge between the CoE and business by leveraging the latest analytics innovations to business and to ensure that the CoE is responsive to real business needs. Given the close collaborations between the CoE and the business, there is enhanced knowledge sharing. With the focus of the CoE on innovations and inputs from business to keep it relevant and actionable, there are continued innovations and the eventual formation of a learning organization. With all these benefits, such a learning and career growth environment is quite attractive to analytics talents, so they result in greater job satisfaction and talent retention.

- **Con**—The downside of the CoE model is that the split control between the centralized CoE and the businesses may be the source of recurring conflicts. Such split controls are byproducts of the intersections inherent in the CoE design. In fact, a successful CoE depends on whether a healthy BAP has been set up for the business and analytics leaders to work closely together. If faced with the complex structure, the leadership fails to get the BAP working properly and lacks the professional maturity to collaborate and closely align on priorities and accountability, potential confusion could arise among the CoE team and ultimately lead to demoralization and talent attrition.

The secret of a successful analytics CoE is how to maintain the proper balance between conflicting interests and demands. In many companies, the CoE model eventually becomes a *de facto* centralized or decentralized model, depending on whether the corporate interests override those of the business units or vice versa.

One possible solution is that the *funding and size* of the CoE should be determined by its success in supporting business. Therefore, the business from its P&L should help to fund its own supporting units and pay a poll tax supporting the corporate CoE. However, the analytics leadership within the CoE should handle the hiring and management of talents with inputs from the respective business leaders. This way, the best talents can be recruited and retained, and synergy is gained through being one team. In short, the *accountability* should be *decentralized*, and the *knowledge and team management* should be *centralized*.

Reporting Structures

Again, there are various CoE reporting structures. It has been common for the head of the CoE to report either to the VP of marketing or the VP of IT to assist marketing efforts. It is also becoming more popular recently to have a VP of analytics (or the chief analytics officer [CAO]) reporting to the CEO/president. Because IT performance is measured differently from analytics, an IT-owned analytics team may result in suboptimal outcomes. Analytics is used to create new opportunities in growing either the top or the bottom line, whereas IT is used to ensure a reliable, problem-free information system at the lowest possible cost.

When applying analytics to businesses, the most ideal model is to have the analytics head report to the most senior person who is tasked with P&L responsibilities for driving the enterprise goals of the company. If growth is the strategic focus, analytics should report to the chief growth officer (if there is such an office or anyone tasked with this role, regardless of title). At eBay, the analytics team reported to the CFO for a while because he was assigned by the CEO to grow (or rather re-grow) eBay's business. Although having a CAO report to the CEO/president may seem like an ideal situation, the premature creation of such a role can be highly risky for the CAO and potentially disruptive for the company. Let me explain.

Because analytics is a powerful change agent that impacts all business silos, it is inevitable that its applications may upset the proverbial apple cart. Unless the CEOs/presidents are confident and secure in their position and willing to lend their bully pulpit to educate teams, analysts, and investors, CAOs are quickly perceived as ineffectual; ultimately, the role is rendered irrelevant and untenable.

Roles and Responsibilities

To build a high-impact team regardless of whether it consists entirely of internal resources or is partly or wholly outsourced, the analytics team must have the right combination of talents performing the following roles. Before I get into the roles, it is worth mentioning that like all high-value strategic assets and activities, the roles that form the core competencies and are central to a company's success, a company should never outsource such critical analytics roles.

Analytics Roles

The roles are roughly split into three main areas: analytics, technology, and business. It is relatively easy to find talents within each area, but the challenge is to find talents that possess skills and experience in more than one area (that is, analytics deciders in the intersections of the Venn diagram shown in Figure 11-1). This is critical because (as discussed in the analytics process) a successful analytics process entails starting with business problem definition, to analytics-formulated solutions, to data audit by IT, to analytics modeling, to insight discovery, and end with business execution. To minimize the loss in translation crossing from one area to another, members with straddling skills are of critical importance and in short supply.

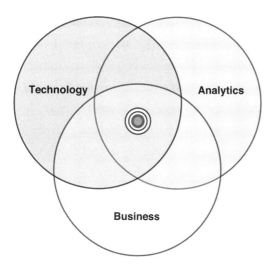

Figure 11-1   Bulls-eye of analytics

Let's look at the roles in detail.

Business Strategy and Leadership Roles

The chief decision/data scientist should be one who has risen through either the quantitative or the business side of the analytics discipline. After years of senior leadership experience, the chief decision scientist should demonstrate an ability to leverage analytics for business success and to be a senior and respected member of the team.

An effective chief decision scientist must be one who appreciates the power of analytics and has hands-on experience in advanced analytics, to be willing to demonstrate passions in business and with gravitas, and to earn and retain trust from senior executive stakeholders. The chief scientist must also function as a strategic member of the senior leadership team chaired by the CEO/president. To promote the cause of analytics, the chief scientist must also lead the team to formulate a consistent and insightful analytics success story that is based on solid scientific discovery assembled from the team's overall outputs.

To the analytics team, the chief scientist must be a visionary leader who can coach, inspire, energize, and guide the team to consistently strive to win and to innovate new solutions. To grow the analytics competency, the chief scientist must devote enough time to develop, recognize, and reward analytics talents as critical assets across key functions of the enterprise. To business clients and sponsors, the chief scientist must be attuned to changing business needs and challenges, and constantly challenge the team to search for new and better analytics solutions.

The chief data scientist or someone familiar with data also performs the role of the data czar. As discussed before, the success of analytics depends on a strong data governance policy. Most companies tend to handle data in an *ad hoc* manner that results in inconsistent policies and standards. In data governance initiatives in many companies, the single most important success factor is the appointment of a strong data governance compliance officer, or data czar. Data governance, like dental hygiene, is not sexy but essential. The data czar must secure CEO sponsorship. Data governance usually elicits positive support, but tends to generate resistance when being implemented. So for data czars to be taken seriously, they must have the backing of the CEO/president when faced with business unit and IT pushbacks. Besides the CEO backing, the data czar must also possess an individual organizational influence (CEO support alone is not sufficient). The data czar must also be respected by everyone in the company and be familiar with business conditions, operational processes, and IT system and data requirements. Instead of compelling compliance, the data czar must use the power of persuasion to convince the leadership that data governance compliance is a win-win for all. The responsibility is to pull together cost analyses for the business on requested changes in operational processes and systems together with IT. To minimize potential conflicts and distractions, the data czar must also possess political savvy and understand the political landscape, be able to build alliances, and make it easier for the leadership to adopt data governance. The data czar must also know how to use effective political levers to ensure continued funding and support from business leaders.

Data governance touches many (if not all) parts of the business. To ensure that important aspects of data governance are always done correctly, a clear data governance process should be established and communicated to all stakeholders. The data czar is responsible to establish and lead an *ad hoc* cross-functional team that meets on a regular basis (monthly) to discuss data quality challenges and needs. The cross-functional team consists of business leaders, analytics, IT, marketing, and sales. In addition, a small support team of privacy experts as well as the privacy office support the data czar. To overcome potential pushbacks, the data czar must have the wherewithal to ensure that the larger enterprise benefits are resolutely defended and not succumb to political maneuvers and pressure. As often seen in some companies, initial interests in data governance may wane over time. It is the job of the data czar to ensure that data governance remains front and center to the business.

Analytics strategists focus on the day-to-day progress of the analytics insights discovery process and serve as custodians of analytics knowledge and its incorporation into the business strategy and executions. They are the ones who provide the link between business and analytics. To do this well, analytics strategists must possess the requisite hands-on analytics skills in terms of fundamental predictive modeling, statistical and mathematical problem-solving methodologies, and DOE. This is important because any successful strategist must be able to communicate effectively with deep quants and also with business leaders. Analytic strategists' skills may also include the following:

- Using advanced and emerging analytics methodologies such as survival analysis; Bayesian statistics; genetic optimization; text mining; semantics analytics; social, influencer, and sentiment analytics; Big Data analytics

- Being able to link business objectives to innovative analytics practices, to extract novel and significant insights from analytics results, and to communicate clearly such insights with business owners in business terms through storytelling and visualizations

- Smoothly implementing analytics recommendations in the field, reporting problems in execution or deficiencies to business and analytics leaders, and suggesting solutions

- Formulating actionable strategies together with the decision scientists

- Defining effective metrics to measure scientifically real business outcomes of the proposed strategies, projects, and team performance in driving business results

- Imbuing the analytics team with a passion for business and strategic thinking

- Educating the analytics team with the latest business focus and the business leaders with new vital analytics assets and acquiring an understanding of what analytics could do for business

Business analysts' skills include the following:

- Supporting the business field team with reporting and analyses work

- Together with the field sales team, gathering input and presenting program results to clients

- Working with the centralized analytics strategist on business requirements for analytics

- Giving the business users an understanding of advanced analytics features and how they may be applied to the respective business

Data and Quantitative Roles

Data/decision scientists should form the backbone of the analytics team because they are the analytics subject matter experts. To do this, they should possess analytics capabilities similar to the modelers, but also focus on the following:

- Understanding more advanced and emerging analytics such as survival analysis; artificial neural net (ANN); Bayesian statistics; genetic optimization; text mining; natural language processing (NLP); semantics analytics; social, influencer, and sentiment analytics; and Big Data analytics (Hadoop and Mahout)

- Translating business problems into analytics briefs and being able to instruct the modelers on the requisite analytics steps, including the required precision and any quality assurance measures

- Being able to maintain a line of sight to the business big picture and being passionate about winning

- Being effective thought leaders in facilitating learning and knowledge transfer as experts in their own subject matters

- Finding novel actionable analytics solutions that others may miss from analytical exercises

- Formulating winning strategies that can be tested and implemented in phases with significant and demonstrable returns along the way

- Being effective mentors to junior members of the team and exhibiting leadership styles consistent with the value and conditions of the team

- Being insistent on scientific measures of performance while being realistic about the need for the business to move forward, and provide realistic guidance derived from partially tested analytics results

The role of *statistical/data mining modeler* should be performed by a statistician or computer scientist with proficiency in machine learning and modern data mining techniques. These modelers should be the ones conducting the heavy lifting of number crunching for the team. To do this, they should possess the requisite analytics skills in terms of fundamental predictive modeling, statistical and mathematical problem-solving methodologies, and DOE by using current requisite tools such as KNIME, R, SAS, Tableau, and SQL. They must also understand the link between business objectives and analytics and always maintain a keen interest in business and the application of analytics. The difference between data scientists and analytics strategists is a matter of focus: Although interested in business, data scientists are not as heavily involved in the day-to-day execution and monitoring of business strategies.

Statistical/data mining modeler skills include the following:

- Interpreting analytics outputs and casting them in clear and demonstrable results
- Communicating results clearly with analytics team members
- Assist the analytics strategists' efforts to formulate actionable strategies
- Assist the data scientist in defining effective metrics to scientifically measure the outcomes of strategies, projects, and team performance

Analytics/data specialists provide the essential data and reporting supports in terms of the following:

- Leveraging the analytics database and sandbox
- Ensuring that the analytics teams have all the data they need to perform their requisite tasks
- Reporting to the analytics team leader

- Performing data exploration tasks that conform to the analytics data requirements

- Using all visualization tools to enable exploratory data analysis (EDA) that may use data stored in the data warehouse or Hadoop and analyzed through Amazon Redshift (or any other chosen technology) and Tableau

- Creating accurate queries that require complex joints with SQL and multiple databases that may be RDBMS, HDSF, or NoSQL

- Constantly transferring essential data and knowledge to the decision scientists and strategists

- Extracting meaningful information from syndicated data sources to support business needs by using the following example for retail and CPG companies. Other industries may have their own syndicated data sources. The basic role, however, remains the same.

  - Nielsen Scantrack (Nitro) for retail and CPG: conduct pricing, promotion, assortment, product performance, and rest of market analyses

  - Nielsen Spectra to assess information to determine opportunities specific to trade areas and stores to determine the best stores for an event program

  - Nielsen HomeScan to extract cross-shopping behavior insights

  - MRI to pull information to better understand the consumer profile for a particular product segment as well as the demographics and psychographics

  - Using Netbase to mine blogs, microblogs, news sites, and chat forums to determine the positive and negative comments that are being shared about retailer, brand, and industry topics

- Assisting the analytics strategists' effort in pulling data to support business growth

Analytics Ecosystem

Given the diverse analytics skills and backgrounds that are also rapidly changing, no single company, regardless of its size, can afford to be an island of analytics excellence. Analytics-mature companies usually bring high-value activities that are core to their business in house and rely on external resources for contingent or non-core requirements. Such external resources, together with the internal analytics team, form the analytics ecosystem, which can include the following.

The In-House IT Function

The IT function, which continues to be responsible for the following:

- **Data collection**—Collecting all relevant data from internal systems and third-party sources; both structured and unstructured data produced from the many emerging media: social, mobile, IoT, or others.

- **ETL process for the deployment phase:**
  - Extracting data from outside sources
  - Transforming it to fit operational and analytics needs
  - Loading it into the database

- **Data architecture**—Managing policies, rules, or standards that govern which data is collected; and how it is stored, arranged, integrated, and put to use in data systems.

- **Data security**—Protecting the database from destructive forces and the unwanted actions of unauthorized user.

- **Data access management**—Managing who in the organization has access to which data sources and information.

- **Data release**—Releasing data to the analytics team.

- **System management**—Managing all IT and analytics systems.

With the increasing amount of data that business collects, it is important that IT continues to focus on these important tasks while the analytics team is responsible for the use and value extraction of the enterprise data assets.

External Analytics Advisory and Consulting Resources

In their quest for analytics competencies, many companies are reluctant to invest in building a large analytics team in the early stages of the quest. The risks of hiring the wrong people can be quite costly and disruptive. It is far better to hire experienced consultants who have demonstrated their skills in building similar analytics competencies in the past and can also fill in for critical roles before hiring full time staff for these roles.

By working closely with the internal business teams, such consultants can also help to specify the requirements, and find and vet the candidates, to ensure that the best people are hired for the roles. In addition, senior consultants who have worked across different industries and problems can bring a fresh perspective to the business team. They also can foster a culture of sharing and help to circumvent the emergence of the not-invented-here mentality that is common in some well-established analytics teams.

How to Choose the Right Consultants

Because it is hard to hire and costly to build up a team of analytics talents, there is always a need for companies to hire external consultants. Even for the largest companies, starting an analytics endeavor can be risky, and hiring the wrong leader may doom the entire project. You might also want to bring in outside consultants to help you bridge the gaps in your current expertise and experience.

Here are things to look for when choosing the right analytics consultants:

- **Hire an analytics decider**—There is always an advantage in procuring the service of a consultant who is an analytics decider because he or she can recognize and hire other analytics deciders. Even though you do not need to staff your entire team with analytics deciders, the early hires should be analytics deciders or have the potential to become analytics deciders quickly. They attract other like-minded talents and can help to establish the team culture. (You may want to refer to the section on how to recognize an analytics decider.) A former chief analytics officer with traits of an analytics decider and a successful track record of planning and building a high-impact analytics team can be helpful in putting together the strategic analytics plan and assist in talent search.

- **Look for prior relevant experience**—Many consultants can claim to have deep experience in analytics, but few have actually worked on the breadth of advanced analytics, especially in a senior leadership position. Many self-proclaimed analytics leaders have neither the expertise nor the mindset of being analytics deciders. Simply looking at the CVs for prior analytics leadership positions in well-known companies does not always indicate competency in analytics. Few companies to-date (including some well-known ones) have progressed beyond analytics support of campaigns and targeting. Moreover, the leaders are usually selected from existing executives who have little or no direct analytics experience. Such consultants cannot help you much because an analytics consultant must have done the things you are planning to implement for your company. You need to make sure that the consultant has dealt with the challenges you are likely to face and has implemented similar and successful solutions.

- **Ensure understanding of your business**—All relevant analytics experiences are necessary, but may not be adequate because every industry and company size may pose different types of challenges. You are better served by making sure that the consultant has a good understanding of your business and how analytics may or may not solve your problems. Similar to choosing a good attorney, those who never mention the weakness of your case most likely are not the ones who will win cases for you. A good understanding means that a consultant can dissect the risks and propose ways to mitigate risks and customize a BAP plan fitting your needs.

- **Access a team of analytics deciders**—Although many vendors can provide modeling resources, few can maintain a team of high-impact analytics strategists that can help elevate your company's analytics competency and find innovative analytics solutions to boost your business results. It is worthwhile to search and procure the service of such consultants, not just for projects but to also provide constant flow of fresh new ideas and perspectives to your business and your internal team. You need to carefully confirm that the solutions provided by the consultants are based on real customer data, follow the BAP methodology, and are fully actionable by your teams within your operational constraints.

Analytics Quality Assurance

Leading external analytics experts should be brought in periodically to benchmark and provide an objective assessment of the business' analytics competency and maturity, with recommendations for improvement.

Affiliations with Colleges and Universities

Given the active research on analytics in both theory and applications, it is recommended that businesses keep close relationships with key universities for the following reasons:

- Leading academia may be engaged in developing new thinking around analytics and R&D efforts to develop innovative business solutions.

- Access to talent pools of qualified analytics talents as interns or permanent hires.

- Promotion of the business' reputation as a leader in applied analytics.

- Participation in key university analytics functions for networking opportunities to learn the most recent best practices in your own and other industries.

- Partnership with universities on important but challenging analytical projects that simply would be too expensive or take too long to do within a business analytics team.

Research and Analytics Trade Organizations

Besides universities, there are many other relevant analytics research and trade organizations that may be useful for business to partner:

- Marketing Science Institute[4]
- International Institute of Analytics Retail Research Council[5]
- LinkedIn Analytics Executives Network[6]
- INFORMS Executive Forum[7]

4 http://www.msi.org
5 http://iianalytics.com
6 http://www.linkedin.com/groups/Analytics-Executives-Network-3007847/about
7 https://www.informs.org/Blogs/E-News-Blog/Executive-Forum-for-Senior-Business-Executives-Strategic-Use-of-Analytics-O.R

- National Center of Data-driven Marketers (NCDM)[8]

Industry Awards

To position your company and be recognized as a data driven and innovative analytics business with cutting-edge analytical capabilities, your business analytics team should consider applying for the following awards as a way to benchmark its competency against its peers:

- DMA/NCDM Excellence in Marketing Analytics[9]
- DMA Innovations Award in Data, Analytics, & Attributions[10]
- ANA Marketing Analytics Leadership Award[11]
- Gartner BI & Analytics Excellence Award[12]
- INFORMS Prize in Advanced Analytics and OR[13]
- MT (Management Today) & Accenture Analytics in Action Awards[14]
- Teradata EPIC Award for Analytics Excellence[15]
- Innovation Enterprise Analytics Award[16]
- TDWI Best Practices Award—Analytics Award[17]
- CITE—Smart Analytics Project[18]

---

8 http://ncdm.thedma.org
9 http://www.chiefmarketer.com/database-marketing/loyalty-crm/the-ncdm-data-base-excellence-award-winners-01022009
10 http://innovationawards.thedma.org
11 http://www.analyticsaward.com/details
12 http://www.gartner.com/technology/summits/na/business-intelligence/excel-lence-awards.jsp
13 https://www.informs.org/Recognize-Excellence/INFORMS-Prizes-Awards
14 http://www.managementtoday.co.uk/go/analytics
15 http://teradataepicawards.com/customer-awards/#analytical
16 http://theinnovationenterprise.com/summits/business-analytics-las-vegas-2014/awards
17 http://tdwi.org/pages/best-practices-awards/recognition-for-business-intelli-gence-data-warehousing-implementation.aspx
18 http://www.citeconference.com/ehome/index.php?eventid=78152&

Analytics Talent Management

As the demands for analytics talents surge, so do the costs of recruiting and retaining such talents. To ensure that the best talents are hired and retained, an analytics organization must be a great and rewarding place to work. Even though some industries (such as retail) are low-margin industries, analytics must still be recognized as a high-value asset and managed accordingly.

More and more businesses are starting to realize that analytics can be used to bring in new revenues without incurring substantial incremental costs and to reduce costs. Given the higher margins, companies can then invest more in building their analytics competency. To do this right, businesses must adopt the best practices in human capital management:

- **Competitive compensations**—Because the business competes with others for analytics talents, it needs to offer competitive compensations. Although it may be harder for businesses in lower-margin industries such as retail grocery businesses, they should benchmark compensations frequently with peers in the same industries (or even with other industries as the best retail analytics talents may be from the hospitality industry) to keep abreast of the compensation changes. When the finance industry first adopted analytics, companies such as American Express had to resort to retention packages known as "golden handcuffs" to stop their top analytics talents from leaving for their competitors.

- **Adequate funding**—Analytics is not built in a day. The multi-year project must be adequately funded and committed to for the long haul. A half-funded initiative is a waste of resources because the initial investment is to build the analytics infrastructure and form the team. To run out of funds when the system and team are finally ready for deployment is a disaster in the making that should be avoided.

Accountability and Timely Recognition

Depending on the particular phase, the team is organized as a matrix organization (that is, split accountability between the corporate and business unit priorities). The corporate and business unit stakeholders also jointly evaluate the performance of the analytics team. A series of awards should be set up for recognizing significant achievements in analytics:

- Innovations in data processing
- Innovations in analytics solutions and applications
- Excellence in the following:

  - Analytics modeling
  - Project management
  - Client relations
  - Team mentoring
  - Business impacts
  - Leadership

Continuous Training Opportunities

The team should be encouraged to develop its own specialties and to share with all team members. Besides informal sharing, learning situations such as the following may be set up:

- **Regular internal seminars**—Breakfast or brown bag lunches where members can share teaching responsibilities should be promoted.
- **Social learning networks**—Blogs, YouTube, and other more recent collaboration tools (for example, jive software,[19] Podio,[20]

---

19 http://www.jivesoftware.com
20 https://www.podio.com

and TOPYX[21]) may be used to advance learning and highlight knowledge gaps and needs.

- **Conference attendance and training opportunities**—To ensure that the team stays abreast of recent developments in the science and application of analytics, the department should maintain budgets (typically 10–15 percent) for professional conference attendance, speaking engagements, and regular training for team members. It has been consistently observed that training opportunities rank highest as the main factor for job satisfaction among analytics professionals.

Career Developments

Recognizing the success of an analytics department depends on team members' diverse skills, so different career tracks should be set up:

- **Separate career tracks**—This is a prerequisite for retaining those who want to specialize as technical specialists versus those who may progress into senior business management roles. The challenge of a technical track for smaller companies may be its short career ladder. However, if the career progression may be measured by business impacts and innovations, the analytics technical ladder may be akin to the senior scientist track within research laboratories such as the IBM Lab. Instead of publications, analytics technical leaders are evaluated by how much impact the analytics technical leader has inside the company, and the reputation and influence outside the company.

- **Clear skills requirements and expected performance**— Career advancements should be set up for each track. Special attention should be paid to develop and reward analytics

---

21 http://interactyx.com/social-learning-blog/elearning-software-remotely-hosts-intranets/

hybrids, who are analytics deciders proficient in both analytics and business.

Only the larger companies have the size to support twin career tracks of sufficient depths for career advancements, however. The optimum for most companies is to have a technical analytics track (in close partnership with external analytics consultants) and use the existing business executive track for those rare talents who are hybrids that possess both analytic prowess and business acumen.

- **Given Line Responsibilities**—One caveat is that such analytics talents should be given line responsibilities instead of conventional staff roles. Real business success is the fuel that keeps the fire of analytics innovations and passions burning high. This is exactly what P&G adopts for the top analytics talents by rotating them to be adjutants to all the business units' general managers (both U.S. and international). Such a rotation is usually reserved for career tracks of rising stars and future CEOs.

Effective Culture and Morale Management

The senior leadership must address how the analytics team fits within the company's culture. To ensure the ultimate success of the analytics team, the inspiration must eventually come from the top (that is, the CEO/president and the senior executive team). How this is done affects the morale of the team and how the business units view and adopt analytics as a valued service. With analytics being a game changer, the analytics leadership runs into many roadblocks. Without the persistent support of the entire senior leadership, there is little chance for the team to be successful.

Analytics as a Revenue or Profit Center

Unlike other horizontal functions such as HR, finance, marketing, and IT, because analytics is so integral to business decisions and is a major driver for the results, it is impossible to treat analytics as a pure cost center. The analytics team should be treated just like sales teams (that is, a revenue center and maybe even a profit center) and given commensurate P&L responsibilities. Ideally, all business owners should be analytics deciders with the support of small teams of analytics strategists that leverage analytics to drive business results. Not many companies are ready for such a bold move today, but given clear and measurable sales targets, there is no problem for companies to attribute P&L to sales executives. The same thing may happen after the contributions of analytics can be measured with the same confidence as sales results.

Conclusion

Building an analytics function is often viewed as a technical endeavor (for example, finding the best statisticians and modelers). However, this chapter shows that for analytics to succeed, it needs the right types of talents, organization structure, affiliated ecosystem, and environment for development and growth. No single structure fits all situations, and the composition of the analytics team also varies from business to business. However, for small- to medium-size companies, it may be best to start with a centralized model and add dedicated resources for supporting different business units over time. This is a phased approach to a CoE model.

As for roles, even though there are different roles, everyone should (in varying degrees) know the three critical success factors: analytics, data/IT, and business. Detailed roles within the analtyics team have also been described. From my experience, I think an ideal composition of an analytics team is 20 percent leadership, 30 percent strategists, 30 percent modelers, and 20 percent data analysts. Because a true analytics decider may fulfill more than one role, the headcounts for each category may vary. Even in a large team, ongoing relationships with external consultants and university affiliates should be maintained to prevent stagnation of ideas and set ways of doing and seeing things. Tips are offered on how to find and select the experienced consultants who are analytics deciders. Once the team is formed, continuous skills training, participation in external meetings, and winning awards and timely recognitions are crucial elements of successful talent management and retention.

12

Conclusions and Now What?

"If we do meet again, we'll smile indeed; If not, 'tis true this parting was well made."

—William Shakespeare, *Julius Caesar*, Act. V Scene 1

In this chapter, I retrace the critical lessons learned and suggest next steps as a part of a strategic analytics roadmap for you and your business. You might want to select and prioritize the recommended steps for implementations.

Analytics Is Not a Fad

I hope you are now convinced that analytics, defined as the "data refinery" in this book, is here to stay. In fact, analytics will continue to grow with the advent of new technologies and scientific methodologies.

Analytics is a true game changer and paradigm shifter. Given today's flood of data and the interconnection of people and things, more novel uses of analytics and data will continue to emerge for decades to come. Today's analytics is often manual and needs customization for each project, which is akin to the age of craftsman preceding the industrial age. In the future, intelligence from analytics is likely to be distributed, embedded, and highly adaptive to each situation without much human intervention. Today, much knowledge exploration still needs to be done manually for each business case.

To-Do: Adopt long-term analytics strategies and invest accordingly. Adopt the business analytics process (BAP) and invest in people and infrastructure.

Acquire Rich and Effective Data

How well analytics can perform depends on how rich and current the data sources are and how well they can be integrated. The power of analytics dictates that data must be as diverse as possible and properly attributed to the same visitor. Data sources should cover different customer characteristics (for example, behaviors, demographics, attitude, preferences, transactions across as many products as possible, and the attributes of products and online and mobile contents accessed). The customer data flow must be connected, aligned, and continuous. To know what effects that a certain mobile browsing data would presage certain customer behavior, the customer behavior data must be first identified as belonging to the same customer and in the same time duration. For causality, the IVs must precede the outcome to be predicted. Before any serious modeling exercises are to begin, the analytics team must do extensive data audit, cleansing, integration, and exploration. Much insight can be gained from such exploratory data analysis (EDA), which often dictates what kind of advanced analytics ensue.

To-Do: Adopt consistent data strategy and invest in acquiring, cleansing and searching, and integrating with new data sources. Establish a data council to oversee data governance and formulate progressive privacy and security policies.

Start with EDA and BI Analysis

Well-conducted initial slicing and dicing of exploratory data analysis (EDA) and business intelligence (BI) visualization tend to raise more questions than answers. This is an important point for business owners: If the EDA and BI investments alone provide all the answers by themselves, the owners are not asking the right questions. The right questions are usually beyond the capability of BI analysis to answer, so the use of advanced analytics techniques is warranted. Conducting advanced analytics without answers to such business questions can lead to wasting precious resources and time. However, without advanced analytics, such questions remain unanswered or (worse) half-baked answers that could lead to suboptimal business results.

To-Do: Invest in the people and BI system for extensive BI visualization and EDA tools. Ensure that the BI and EDA process produces the right business questions for advanced analytics and for knowledge extraction.

Gain Firsthand Analytics Experience

For business owners to ask the right questions, they must go beyond just knowing the BI and data analysis, and progress to the realm of advanced analytics. Without knowing advanced analytics, it would be hard, if not impossible, for business owners to fully comprehend, instruct the team, and leverage the power of analytics in their business. This is why this book takes on the unusual task of trying to help business owners acquire hands-on experience.

To-Do: Gain firsthand analytics experience. Maintain a fresh interest and curiosity about analytics and constantly try to learn, regardless of how long you have worked and how senior you may be in the company organization.

Become an Analytics Decider and Recruit Others

Business owners who want such hands-on experience are probably analytics deciders, who thrive within the intersections in which different talents need to fully collaborate and work closely together. Companies desirous of constant innovations must endeavor to create rather than eliminate such collisions within the intersections. BAP is built specifically around such intersections. The success of companies becoming analytically competitive depend on whether they 1) have adopted the BAP methodology, 2) have well-established and fully functioning intersections along the critical points within BAP, and 3) have an adequate number of analytics deciders who are happy and thriving within the intersections.

Any analytically competitive business should be a haven for Medici effects to occur. Any business or technology professionals aspiring to be analytics deciders can start from where they are and follow the suggestions to acquire the requisite skills to thrive within the intersection. Tips are also given for businesses to find, grow, reward, and retain the analytics deciders.

To-Do: Become an analytics decider. For your business, set up the BAP infrastructure and create safe and functioning intersections for the Medici effect to occur. Find, recruit, grow, and retain adequate numbers of analytics deciders for your business to win.

Empower Enterprise Business Processes with Analytics

In the perspective of analytics, the enterprise business process should be upgraded. Unfortunately, the current Enterprise Resource Planning (ERP) processes are often hard-wired and cumbersome for supporting such *ad hoc* analytics. It is proposed that either a connector or bridge be built to enable analytics to run directly on the data within the ERP system.

Another way is to store all the raw data output from the ERP within a Big Data NoSQL environment for exploration and analytics knowledge extraction. The beauty of this is the untethering of the analytics from the cumbersome ERP system. Within the analytics sandbox, analysts can run the analytics and the extract, load, and transform (ELT) processes on demand and in a matter of minutes without waiting for the ERP system to crunch through the requests. The analytics solutions can then be embedded and optimized to run directly on the database.

To-Do: Leverage all the data assembled through the ERP to enable analytics to add intelligence and insights. Set up ERP connectors or an analytics sandbox to enable ELT and analytics on demand.

Recognize Patterns with Analytics

Similar to problems in nature, business problems exhibit patterns, and the ability to recognize these patterns often leads to significant insights and winning solutions. Even though it is possible to render customer relationship management (CRM) in a one-to-one manner, the strategies and levers need to be grouped and tested for specific customer segments. Rather than start with a preconceived segmentation scheme, it is often much more effective to use unsupervised learning to cluster the customers. The clustering technique attempts to produce segments with the lowest degree of inhomogeneity. Segments with a high degree of inhomogeneity mean wasted resources and lower conversion rates and returns on investment (ROIs).

Not only customers exhibit patterns; products purchased by customers tend to also exhibit certain patterns. By understanding how products are bought together, effective promotions and product display layouts can be used to encourage customers to buy more products. Affinity or market basket analysis can be used to determine which patterns have the highest support and produce the greatest lifts. If the customers can be identified with POS receipts, different patterns may be prescribed for the different segments. By combining the patterns with the respective segment personas, the reasons or utilities that underlie the purchase decisions per segment may be surmised and validated, perhaps through carefully designed customer surveys.

History tends to repeat itself. Leveraging this, past histories have been successfully used to predict future events. The history may be from its own past or from related entities. After such patterns are found, various levers can then be tested to determine causality and optimal lever setting. From these, scenarios may be constructed to simulate the outcomes and marketing ROIs under varying conditions. The ability to predict future outcomes also shows emerging trends for appropriate provisory measures.

To-Do: Hire and train analytics talents who can recognize patterns in customers and product purchases. Leverage past histories to predict future outcomes and trends. Identify and optimize levers (promotions, sales, new products) in scenario simulations and predict levels of investments and ROIs.

Know the Unknowable

With today's data and advanced analytics tools, few things in business are truly unknowable. When an event first appears as unknowable, the reaction should not be that it can't be done. Instead, the right approach is to follow the BAP to audit the data and see whether any analytics tools can be combined in a novel fashion to shed light on how to analyze and predict the event. One critical missing piece in most business planning processes is the actual customer wallets and their total spent over the customers' purchase life cycle. It was the absence of this critical information that caused most businesses to adopt a vanilla flavor approach to CRM. In their efforts to please every customer, they end up delivering a lower level of customer service to all. By being able to predict the values of its customers and their segment personas, a business can tailor its entire customer engagement strategy. It can choose appropriate strategies based on what would appeal most to the customers per segment and to allocate investments commensurate with the customers' respective values.

These predicted values and the effectiveness of the levers should be properly validated using scientific test protocols such as the design of experiments (DOE). For levers, the capability to establish causality is of paramount importance. Correlations alone are not adequate to ensure that the same levers will produce similar results in the future.

To-Do: Predict customer wallets per product and for the entire business. Predict addressable LTVs per customer (per segment). Predict wallet and wallet share per customer, and identify and estimate the effects of the respective levers to affect wallet share gains. Always use proper scientific multivariate tests to establish causality. If such tests are not possible, use the PSM methodology to establish causal effects by constructing equivalent control groups by their respective propensity of belonging to the test group.

Imbue Business Processes with Analytics

Instead of leveraging analytics for tactical activities such as campaign targeting, you have seen how the application of analytics can pervade every part of the business process. In fact, many novel strategic uses of analytics remain to be discovered. By mastering various tools to work on the data, you might be able to put together simple workflows in which analytics tools can be combined to produce novel results and unique insights. The ten questions may be further customized to generate many other related questions for specific business situations. You are encouraged to try your hand at constructing such workflows for answers to your own questions.

To-Do: Prioritize questions and try to formulate workflows to answer these specific business questions. Ensure that the more advanced predictive and prescriptive analytics are used instead of the simple descriptive and diagnostic analytics.

Acquire Analytics Competencies and Establish Ecosystem

As analytics becomes more and more popular, many businesses are starting to evaluate their analytics competency and maturity. Most businesses may think that acquiring analytics competency is a matter of hiring an analytics team and buying the best tools. However, analytics competency and maturity are more than people and tools; they are also about organization, enterprise leadership and focus, and a full ecosystem of academic affiliates and external expert consultants.

It was the aim of this book to provide a summary from my years of experience in forming high-impact, learning, and passionate analytics teams. Depending on the size and particular maturity state of the team, the business may want to adopt a specific organizational structure. With the close collaboration of business with the analytics leaders (analytics deciders), the initial team structure may then be gradually transformed into a more appropriate model.

You could start either with the centralized (more appropriate for SMBs) or the decentralized (for larger companies with multiple lines of business) model. As long as both the enterprise and business needs are kept in constant balance, the structure will ultimately become, in practice, a Center of Excellence (CoE).

The challenges usually occur when a certain business interest precedes another, and eventually the boundaries of the silos (either between the business units and corporate or among the business units) create either an ivory tower or scattered islands of analytics competency, respectively. It may be beneficial for the internal stakeholders to hire a third-party outsider to ensure the balance, put in appropriate key performance indicators (KPIs), and benchmark progress. An experienced external consultant who has direct experiences in building and managing the process should be an essential part of the ecosystem.

To-Do: Benchmark the business analytics maturity and competency by procuring the service of qualified SMEs. Create analytics roadmaps for achieving analytics competencies. Hire external experienced consultants to ensure the appropriate structure, training, reward, and culture; and affiliate environments for continuous learning and growth.

Epilogue

I hope you enjoyed reading this book and will find it useful as you embark on the journey of business analytics application as an analytics decider. As Cassius said in the opening quote, I hope this parting is indeed well made. If you want to share your story or engage in further discourse, please feel free to email me at clin@sloan.mit.edu.

The Appendix contains materials that are important but are too detailed for inclusion in the main body of the book. This includes the basics on the Analytics tool KNIME and descriptions of some of the KNIME nodes. For ease of reading, some of the passages may be repeated here from the main body of the book.

KNIME Basics

To get started with KNIME, you should go to http://www.knime.org/knime-desktop-sdk-download and download the most recent version of KNIME for the Linux, Mac OS, or Windows version.

You can view a short introduction video at YouTube (https://www.youtube.com/watch?v=ft7Ksgss3Tc) or search key words KNIME introduction on Google.

For a more thorough introduction of KNIME, starting from the very basics and then covering data manipulations, data joiner, basic statistics, regression, and simple clustering, check out the e-book written by the KNIME Guru, Dr. Rosaria Silipo. A sample of the first two chapters can be downloaded for free from http://www.knime.org/files/bl_sample_s.pdf or you can purchase it from http://www.knime.org/knimepress/beginners-luck for €19.95 (or about $26.00 U.S.).

For advanced users, the same author has also written a useful book called *KNIME Cookbook - Recipes for the Advanced User* (https://www.knime.org/knimepress/the-knime-cookbook). It can be purchased for €19.95 (or about $26.00 U.S.).

You are encouraged to download the KNIME tool, and download and consult the introductory documents before viewing the sections that follow. Like cooking, you can only learn by doing the hands-on exercises. So start cooking!

Data Preparation

This section covers issues related to preparing data before a BI analysis or visualization is undertaken. The issues involve:

- The different types of data formats
- Treatment of missing values
- Normalization or standardization of variables
- The use of data partitions

The efforts in cleansing and preparing the data in most analytics projects usually constitute 30 to 40 percent of the time and efforts. The actual percentage depends on the size, state, and complexity of the actual data.

Types of Variable Values

For analytics, data needs to be pre-specified and read according to their types. For example, KNIME *File Reader* node would try to "guess" the data types when reading an input file, and it would not always guess correctly. So additional manipulations are often necessary to ensure the data is read in correctly. There are several types of variable values supported by KNIME: **I** stands for **I**nteger values, **D** for **D**ouble or numerical values, and **S** stands for **S**tring values.

The string values can also be nominal (names), in which case they can be mapped to an integer using the KNIME node *Category to Number*. Similarly, other variable values may be changed from one to another using the nodes *Number to String*, *String to Number*, *String to Date/Time*, or *Time to String*. Date and Time can also be configured in the respective nodes to different date and time formats. The *Column Rename* node may be used to change all variable names and types in one node.

Dummy Variables

Another way to deal with nominal variables is to create the so-called dummy variables where the nominal or category variables are transformed into "dummy" numerical variables of "1" or "0." For example, the gender field when used together with other numerical variables such as age or number of years in school may be transformed to two variables, "Male" and "Female". A male gender would be "1" in "Male" and "0" in Female. By default, if both variables contain "0," then the field is not specified or missing. This may be significant if the gender information is to be provided by customers. A missing value may simply indicate an unwillingness for the customer to provide this field. KNIME node *One2Many* may be used for creating the dummy variables.

Missing Values

Missing values should not always be taken to be zero. For example, an online shopper who browses but does not buy or who has abandoned the shopping cart or deleted an order in the shopping cart are not the same as those shoppers for whom the system did not show any purchases for the same period. Depending on the context surrounding the missing values, the appropriate missing value imputation method should be used. For example, KNIME *Missing Value* node as shown in Figure A-1 has the following options:

- **Do Nothing**—Choose this option if the missing data can safely be ignored without causing any confusion or errors in the workflow.

- **Remove Row**—The missing data indicates the other field values in the same row are of questionable validity, and hence, the entire row will be removed.

- **Min**—The default or baseline value of the entire dataset for that field is taken to be the minimum value.

- **Mean**—The missing data is taken to be the average value for the particular variable.

- **Max**—The default or baseline value of the entire data set for that field is taken to be the maximum value and replaces all missing values.

- **Most Frequent**—The missing data is taken to be the same value as the most frequently occurring values for the particular variable.

- **Fix Value**—The missing values correspond to a certain value; for example, missing discount values for coupons given out during in-store tasting promotions should be taken to be the same as non-redemption, and hence, zero.

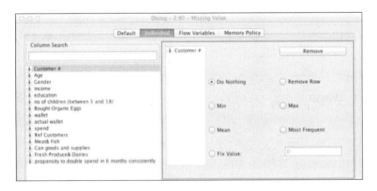

Figure A-1 Missing Value KNIME node

Data Normalization

In many models, the range of the IVs can vary greatly. Some algorithms require the IVs be of similar range, such as in clustering. In most clustering, a distance measure needs to be defined. Without standardization, a small percentage change in an IV with an inordinately large range will have larger impacts on the clustering results than the same percentage of change in IVs that have smaller value ranges.

There are three ways of normalization:

- An absolute range—that is, maximum-minimum
- A standard deviation—that is, Z-score normalization, with Z-score and essentially the number
- A percentile ranking

KNIME provides a convenient way to do this through the node *Normalizer*. In some cases, one needs to de-normalize the values to determine the characteristics of the clusters. In terms of actual values, the node *Denormalizer*, as shown in Figure A-2, may be used.

Figure A-2 KNIME node Normalizer of IVs

Data Partitions

As part of the supervised modeling process, it is typical to divide the dataset into two or three samples known as *partitions*. The first partition is usually used to "train" the model. The second partition is usually known as a *validation partition*. The validation sample is used to validate the trained model without further model adjustments.

Sometimes the second sample is used to choose the best combination of model parameters such as the number of clusters (to prevent over-fitting) or the best model out of several models. In this case, a third sample known as the "test" sample is then used to test the best model without further changes.

KNIME *Partition* node, as shown in Figure A-3, makes the task of partitioning a fairly easy task. Some of the important notes on the use of the Partition node are listed below:

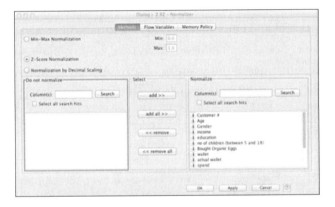

Figure A-3   KNIME Partition node configuration

- The size of the partitions may be selected in two ways:
  - **Absolute**—The number of rows in the first partition.
  - **Relative**—The percentage of the entire dataset to be in the first partition.
- The first partition can be sampled in one of the following ways:
  - **Take from the top**—It will sample from the 1st row until the specified sample size is reached.
  - **Linear sampling (known as the Nth sampling)**—Where a 1% sample that is sampled from the top consists of the set of rows ...1st, 101st, 201st ....

- **Draw randomly**—The KNIME Partition node will draw a sample randomly in two ways. It will draw it without a fixed seed, which means every time you invoke this node, it will draw a different sample. Or, if you want a specific randomly drawn sample that you can return to check the results, then you would specify a certain seed.

- **Stratified sample**—If there are nominal values, such as those shown in Figure A-3, if we check Stratified sampling, it will draw a random sample but keep the distribution of values in the Gender field of the entire population the same. (If there are no nominal variables, then the Stratified sampling will not be selectable.)

When more than two partitions are needed, you can create a single meta-node with the number of *Partition* nodes nested together; that is, three partitions will be a two-nested *Partition* node, and so for n partitions, one would need to nest n-1 nodes together.

Exploratory Data Analysis (EDA)

Instead of exploring a multi-dimensional big data problem, let's simplify it to something we can easily handle and visualize, such as a three-dimensional data cube. A larger dataset would simply mean a cube in more than three dimensions and a much larger number of sub-cubes.

Multi-Dimensional Cube

We illustrate the concepts of EDA in the next section using the example as shown in Figure A-4. Let's assume we are interested in sales of your stores. The store sales in U.S. dollars are arranged by respective products (bread, milk, eggs, meat, and fruit), by their locations (New York City, Boston, and LA), and for the months of January, February, and March.

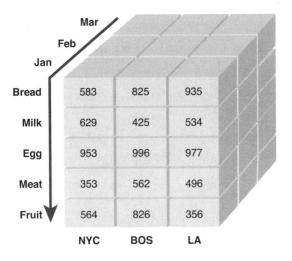

Figure A-4 Data cube

For starters, we can perform some simple statistical character-ization. Again, KNIME gives us a node to do just that. Node *Statistics*, as shown in Figure A-5, computes the Min, Max, Mean, Std. deviation, Variance, Overall sum, and other.

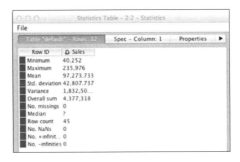

Figure A-5 KNIME node Statistics

In addition to the simple statistics, you may want to see how sales in each store can vary over time, or by products. To answer these and other related questions, you can resort to operations known as slicing, dicing, drill down/drill up, and pivoting.

Slicing

As shown in Figure A-4, one can perform a vertical or a horizontal slice to answer different questions. To get all the sales in January or for the New York City store, one would take a vertical slice along one of the two dimensions. To get the former, the vertical slice will be along the month="JAN" dimension. A horizontal slice would answer the question regarding products. For example, to find out how many meats were sold, we can sum over the sales from the fourth horizontal slice from the top.

Dicing

Dicing isolates a particular combination of parameters. For example, the sales of milk and eggs in the Northeast region for Q1, excluding the January month (atypical number due to unusual weather), will result in a cube of 2x2x2.

Drilling Down or Up

Assume there are subcategories below each dimension. For example, below Month, you may find weeks; below Cities, you may find the store numbers; and below Meats, you may find types of meats, and then you can drill down to the subcategories. Or, one can "drill up" by summing the subcategories into their higher dimensions, such as summing the months into quarters.

Pivoting

Pivoting allows one to answer even more complex questions than just slicing and dicing. For example, if you want to know how the product sales vary for each of the three cities and you do not care how they vary over time, then you can pivot by cities and summing over the sales for the city over the three months. KNIME has a *Pivoting* node that can help do this.

KNIME *Pivoting* node (as shown in Figure A-6) can be configured for groups, pivots, and options (as shown in Figure A-7), as follows:

- **Groups**—This dimension produces a Groups totals when *Pivoting* in KNIME. If the defined Options is set to sum the sales, then the Groups total will give the total sales per product summed over the cities and months.

- **Pivots**—These are the dimensions for which a pivot table is generated. In the previous example, this will be the sales for each city summed over the three months.

- **Options**—For the pivot options, one can choose which column to aggregate and also the Aggregation method. The Aggregation methods available are: Mean, Standard Deviation, Variance, Median, Sum, Product, Range, Geometric mean, Geometric standard deviation, First, Last, Min, Max, Mode, … Unique count, Count, Percent, Missing value count, and so on. The most common aggregation methods are Mean, Sum, Count, and Unique count.

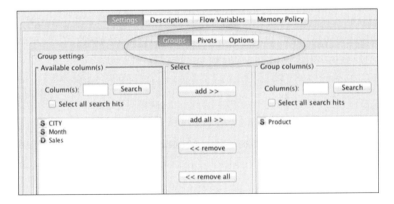

Figure A-6   KNIME Pivoting node configuration

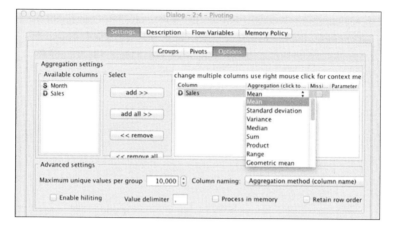

Figure A-7   Pivot options

Index